TENDRILS OF REDEMPTION

ASHLEY JONES

First edition

ISBN: 979-8-218-89957-8

This book was professionally typeset on Reedsy. Find out more at reedsy.com

TABLE OF CONTENTS

One

NIGHTMARES IN THE SKY

The first to sense it were the gulls. They wheeled and shrieked above Haven's Reach, their cries sharper than usual, cutting through the evening hush like blades. Then, without warning, the entire flock broke from the cliffs as one, scattering skyward in a frenzy of wings before vanishing into the horizon. In their wake, the docks lay hushed, a silence so heavy even the waves seemed to hesitate, as if the sea itself had forgotten how to breathe.

On the boardwalk, young Thomas Calder tugged on his father's sleeve, his small fingers clutching with urgency.

"Pa," he whispered, voice thin against the hush, "why's the sky glowing?"

His father squinted into the distance, one calloused hand shading his eyes against the strange shimmer bleeding across the horizon. At first glance, it looked like the northern lights, pale ribbons unfurling, stretching themselves lazily across the heavens. But the colors were wrong, too green, too blue. They bled into one another like oil on water, restless, shifting as though alive. The longer he stared, the less it resembled a light show and the more it looked like something moving, writhing, breathing just beyond reach.

"Storm lights," the fisherman muttered, though his voice wavered. "Get inside, lad."

Down the boardwalk, other fishermen had paused in their work. Nets half-pulled from the water dripped steadily, forgotten in their calloused hands, while every gaze strained toward the horizon. The glow spread slowly, like oil blooming under glass, painting their weathered faces in restless shades of green and blue.

Among them stood Alex Shepherd, a coil of rope slung over one arm, his stance steady as the pier itself. He said nothing, but when the younger men glanced his way, they seemed to settle, as though his silence lent them courage. One, fumbling with the wet net, called out, "Alex, what do you make of it?" The man's voice cracked, but Alex only shook his head, eyes fixed on the strange lights. No bluster, no guesses, just watchfulness. When the net slipped from the boy's hands, Alex caught it with an easy motion, passed it back, and returned to staring out at the sea. His presence alone seemed answer enough: the situation was grave, and it demanded patience, not panic.

Mrs. Callahan, wrapped tight in her shawl on her porch as she was every evening, leaned forward with a frown. The beads of her rosary clicked softly between her fingers, the rhythm of prayer as steady as the tide.

"The sea's unsettled," she murmured, her voice low but sharp enough to cut through the uneasy silence. "Mark my words, there's ill in that light."

She rose slightly from her chair, eyes narrowing at the shifting glow that spread across the horizon. The colors rippled in patterns too deliberate to be chance, and her breath quickened.

"Not storm nor star," she whispered, clutching the crucifix to her chest. "It's the old spirits stirring, the ones who walked the deeps before men ever cast nets in these waters. The sea remembers. It always remembers."

Her words carried down the boardwalk, hushed yet insistent, drawing sidelong glances from the villagers. Some muttered that the old woman had always been touched by superstition, but no one laughed. Not with the air so heavy, not with the gulls fled, not with the strange fire spilling over the waves.

Mrs. Callahan's gaze never wavered. Her fingers worked faster along the rosary, each bead slipping through like a heartbeat. "The gods are listening," she warned. "And when the sea speaks back, it's never for nothing."

A group of children, still chasing one another between the fish crates, stopped mid-game. Their laughter fell away like a cut string. Small hands clutched the wood and rope as they craned their necks upward, eyes wide and shining in the strange light.

One boy swore the stars themselves were moving, tracing patterns that had never belonged to any constellation. Another, his voice trembling but eager, insisted it was angels descending, their wings burning blue and green against the heavens. The youngest, barely old enough to toddle after the rest, buried her face in her brother's sleeve and began to cry, the thin wail sharp in the thickening silence.

"Don't cry," the older boy muttered, though his own voice shook. "It's just... it's just lights." Yet his gaze never dropped from the sky, and his grip on her hand tightened until their knuckles blanched.

One daring girl cupped her hands around her mouth and shouted to the heavens, as if the glowing sky might answer back. Her echo returned hollow, swallowed by the hum beginning to rise, and she fell silent, suddenly unsure.

The children huddled close, the game forgotten. Their whispers mixed with fear and wonder, carrying through the market square like sparks caught in the wind.

Even the tavern door creaked open, spilling lamplight onto the lane as the innkeeper stepped out, wiping his hands on his apron. He followed the children's pointing fingers and froze, the cloth slipping soundlessly from his grasp.

For a long moment, he stood in the doorway, caught between the warm roar of laughter behind him and the cold, shifting glow above. His lips moved, forming words he didn't speak aloud, perhaps a prayer, perhaps a curse. The light painted his weathered face in hues of green and blue, so strange it made him look like a man half-drowned, risen from the depths to stare at the sky.

The tavern's laughter faltered. One by one, the voices inside dimmed until only the crackle of the hearth could be heard. A few patrons leaned into the doorway behind him, mugs clutched in their hands, their eyes wide and glistening as they too looked upward. The music that had been playing, a fiddler's tune to drown the long hours, died mid-bow, the last note quivering into silence.

The innkeeper's apron fluttered in the sea breeze, his broad chest rising and falling with shallow, unsteady breaths. At last, he muttered, rough and low, "That Ain't storm light."

His voice carried into the lane, and those who heard it shivered, because it was not the voice of a man guessing, it was the voice of a man certain.

Across the square, shutters opened one by one, hinges groaning as if reluctant to break the hush. Faces emerged, pale and ghostlike in the shifting glow, their eyes wide and unblinking. Mothers clutched shawls tighter around their shoulders, men leaned out with calloused hands braced on windowsills, and children pressed their foreheads to the glass, breath fogging the panes.

No one spoke above a whisper. A murmur here, a gasp there, as though any louder sound might ripple through the night and draw the gaze of whatever lingered above the sea. Even the usual sounds of Haven's Reach, the clatter of buckets, the bark of dogs, the creak of rope, seemed to wither into silence, as if the town itself dared not breathe too deeply.

The glow painted the square in unearthly hues, shadows bending in unnatural ways, and for a fleeting moment, it felt as though the village had slipped into another world entirely, suspended, waiting, listening.

The air thickened, heavy as wet wool, and every breath seemed harder to draw. At first it was subtle, a low hum pressing against eardrums, so faint most dismissed it as the rush of blood after a long day. But then it deepened, steady, insistent, crawling through marrow and muscle alike. Teeth began to chatter without cold, mugs on windowsills quivered against the wood, and the lantern glass trembled in its frames until flames inside guttered and bent sideways, as though bowing to an unseen current.

The hum was not merely sound. It was a presence.

The nearest lantern burst with a sharp crack, glass raining down across the wharf in glittering fragments. Its flame died with a hiss, smoke curling upward before the wind carried it away. Darkness swallowed that stretch of boardwalk, broken only by the sickly green-blue shimmer rippling above the sea.

People recoiled as though struck themselves, hands flying to ears, jaws clenched against the vibration that seemed to live inside their skulls. Mothers pulled children close, fishermen muttered broken prayers, and even the dogs whimpered low in their throats. The hum pressed harder, relentless now, until it felt as if the entire town had become one great drumhead, strung tight and waiting to split.

A fisherman staggered, his net slipping from his hands in a wet slap against the planks. He clutched his head, eyes wide and bloodshot, before a crimson line spilled from his nose and streaked down his beard.

"It's in my head," he rasped, voice breaking, before crumpling to his knees.

Two mates caught him under the arms, hauling him upright with grim effort. Their own faces gleamed pale with sweat, jaws locked tight as if they, too, were fighting some invisible weight pressing inside their skulls. One spat into the sea, the motion shaky, as though to purge whatever force had invaded them.

The watching crowd shrank back, whispers scattering like leaves in wind. Some crossed themselves, others clutched charms or scraps of driftwood strung around their necks. And still the hum pressed on, relentless, demanding attention like a voice no one could yet understand.

From the stables came the thunder of hooves striking wood, the frantic sound of animals gone wild. Horses screamed and reared, their bodies crashing against stall doors until hinges groaned and splinters flew. Chickens erupted from their coops in a flurry of wings, feathers spinning like pale snow through the lamplight. Even the dogs, once loyal shadows at their masters' heels, snapped their tethers with sharp yelps and bolted into the alleys, tails clamped tight between their legs, vanishing as though chased by unseen predators.

"Not natural," Mrs. Callahan muttered, her rosary clutched so tightly the beads bit into her fingers. She rocked on her porch, shawl pulled close, her eyes fixed on the horizon where the glow spread wider. Her voice, thin but certain, carried to those near enough to hear;

"Not storm, not sky. Something's stirring out there. Something is wrong in the sea."

The way she said it, like a warning carried on the bones of old prayers, sent a ripple of dread down the lane. Even those who scoffed at her mutterings shifted uneasily, as if afraid she might be right.

Along the wharf, the telegraph wire shuddered on its poles, its metal strands quivering like a plucked string. Then, with a sharp crack that split the night, it snapped. Sparks spattered into the dark, showering over the boards before vanishing into the tide. The hum swallowed the sound as though it had been waiting for it, deepening until the silence left behind felt heavier than the sea itself.

The town stood cut off, no messages out, no warnings in. Haven's Reach belonged to the glow and the hum alone.

The water itself shimmered, as if stirred from within. Shapes took form beneath the surface, shadows that pulsed with cold fire, drifting upward, nearer, nearer, until they breached. Translucent bodies unfurled against the heavens, their bells swelling, their wings catching what little starlight remained. Dozens of them rose together, drifting upward in slow, deliberate arcs that mimicked the breathing of the tides. Their glow spilled across the harbor, painting the masts in ghostly green and blue, sending rippling ribbons of light across the faces turned skyward.

Gasps broke from the crowd, sharp but hushed, as though fear of breaking the moment held every throat tight. For an instant, awe silenced Haven's Reach. The creatures moved with such impossible grace that even the gulls circling high above fell quiet, their wings frozen mid-arc as if unwilling to disturb what hovered below them.

The first of the glowing shapes cleared the rooftops, its tendrils trailing like silver fire, swaying in some rhythm older than the sea itself. The air prickled with their radiance. Children clung to their mothers' skirts, eyes wide, while grown men lowered their nets with trembling fingers. It was beautiful, unearthly, and unbearably close.

Then the stillness soured. The tendrils lingered too long above the square, swaying like grasping fingers. The glow that had dazzled the townsfolk shifted, growing harsh in its brilliance, casting shadows that cut sharp across the cobblestones. What had seemed like wings now looked like veils drawn tight, hiding eyes no one could see but everyone could feel.

Mrs. Callahan's rosary slipped from her hands and hit the wood of her porch with a hollow clatter. The sound, so ordinary, so human, rang out against the unearthly silence, and in that instant, the awe cracked. Whispers rippled through the crowd, thin and frightened.

Then, from somewhere in the knot of villagers, a voice cut through the hush:

"Turner's curse."

The words hung heavier than the hum itself, catching in every ear like a hook. A fisherman crossed himself. Another spat into the dirt, muttering about forbidden science and the sins of meddling hands.

Olivia's name was not spoken, but every head turned as one, toward the cliffs, where her laboratory windows gleamed faintly, reflecting the same unnatural light that bled from the sky. The sight of that glasshouse, a place most had long whispered about, now blazed like an accusation.

"Her kind's been tampering with what ought not to be touched," a woman hissed, dragging her child against her skirts. "Now look what's come of it."

Others shushed her, but the whisper was already alive, spreading fast like sparks jumping across dry grass. Awe curdled into suspicion. Fear, searching for something human to blame, settled on Olivia Turner, though she wasn't even present among them.

Near the edge of the crowd, Alex Shepherd shifted. Broad-shouldered, arms folded across his chest, he said nothing, but the set of his jaw was answer enough for anyone who glanced his way. His silence carried authority, the kind that made men lower their eyes and think twice.

Someone muttered too loudly about "witch work," and Alex's head turned, slow and deliberate. His gaze cut through the words like a blade. He didn't raise his voice, he never did, but the look alone silenced the speaker, who cleared his throat and shuffled back into the knot of villagers.

Alex remained still after that, a shadow against the lamplight, but every line of him radiated warning. He knew this town too well, its quick judgments, its hunger for scapegoats. Haven's Reach needed no push to turn suspicion into condemnation, and though he would never speak the words aloud, his stance made it clear: if their fear sharpened into accusation, they would have to pass through him first.

The hum pressed harder, rattling glass, but in that square, it was the hush around Alex that carried the most power.

Above them, the jellyfish pulsed in perfect unison, as though answering the accusation. The timing was too sharp, too deliberate, as if the creatures themselves understood the weight of the words spoken below.

They stared, breath caught in their throats, as the glowing forms drifted higher, tendrils trailing down like ribbons of pale flame. Each pulse of light was synchronized, each beat echoing in the bones of Haven's Reach. The sound was no longer a hum but a drum, steady and unstoppable, the heartbeat of something vast and alien.

Then one tendril, thin as a whip of glass, sagged lower than the rest. It swayed lazily above the rooftops before dipping further, brushing the eaves of the Calder house.

The contact was instant and brutal. A hiss, sharp as searing fat on a skillet, ripped through the silence. The shingles blackened, edges curling inward, smoke twisting skyward in a ghostly spiral before the whole patch crumbled into a spray of glowing ash. Sparks floated like dying fireflies across the crowd.

Children shrieked. Men surged forward with buckets, though there was no fire to douse, only the scar of what had been burned away without flame. Women clutched their rosaries tighter.

Mrs. Callahan's voice carried through the chaos, raw and unsteady:

"Spirits preserve us. It's a scourge, a judgment sent from the deep!"

And still, above it all, the jellyfish pulsed in unbroken harmony, their tendrils trailing lazily, as if reminding the town how fragile every timber, every rooftop, every life truly was.

A woman screamed, the sound cracking through the square like a gull's cry. Men surged forward with buckets drawn from the well, sloshing water over the Calder roof in frantic waves. Steam hissed where it met the scorched timber, though there were no flames to chase, only the smoking scar left by the brush of an alien tendril.

Then another tendril descended, swaying low across the boardwalk.

It grazed the planks with a slow, deliberate sweep. The wood hissed and warped, swelling outward before caving in, glowing faintly as if fire licked it from within. The smell of brine and charred timber bled into the salt air, sharp and choking.

Children wailed. Someone shouted for the bucket line to move toward the docks, but the villagers froze, paralyzed by the sight of the boardwalk glowing as if alive, each scorched groove pulsing faintly in rhythm with the jellyfish above.

The creatures didn't hurry. They drifted higher again, languid, unconcerned, as if showing how easily Haven's Reach could be unmade, plank by plank, roof by roof, whenever they chose.

The church bell tolled wildly, its rope yanked by a terrified boy. The frantic clangor should have split the night, should have carried across every rooftop, but even its peals bent and faltered beneath the drone, swallowed until nothing remained but the endless, inescapable hum.

That sound broke Haven's Reach.

Mothers clutched their children so tightly the little ones whimpered, dragging them into doorways and bolting shutters behind them. Men slammed bars across windows, shoulders hunched as though the very air were pressing them down. The tavern emptied in a stumbling rush, mugs and dice scattering across the floorboards as the keeper shoved out his last patrons and barred the door with trembling hands.

In the stables, horses reared against their reins, hooves clanging against wood, eyes rolling white with terror. Chickens burst into the street in a frenzy of wings and feathers, scattering into shadows as the hum made the very earth vibrate.

"Inside! All of you, inside!" a man bellowed, voice hoarse with strain, but the words scattered like dry leaves before a gale. Panic had already taken hold. The crowd was beyond listening. They fled in every direction, shouting, stumbling, hands pressed to their ears as if they could block the resonance crawling into their bones.

But not all moved.

At the steps of the church, Alex Shepherd stood unmoving, broad shoulders squared, arms crossed loosely over his chest. The glow from above painted his face in shifting shades of blue and green, but his expression didn't waver. While others bolted, Alex remained as steady as a mooring post in a storm. His silence carried more weight than the shouts. A few men glanced his way mid-flight, shame flickering in their eyes at their own panic before they vanished into doorways.

He said nothing. He didn't need to. His very stillness steadied the space around him, even as the swarm blazed brighter overhead.

The jellyfish pulsed in perfect unison, their tendrils hanging like pale banners. Each synchronized flare washed the cobbles, the rooftops, and the cliff face itself. Too steady. Too deliberate. A pattern. A rhythm.

A signal.

And though no one dared speak it aloud, every soul who cowered in the dark knew the same terrible truth: the creatures were not merely drifting overhead, they were communicating, perhaps to each other, perhaps to the deep, and Haven's Reach was caught beneath their message.

Alex finally lifted his gaze skyward, jaw tight, eyes narrowed against the light. He didn't flinch. He didn't run. And though he spoke no words, in that moment anyone watching would have known: Alex Shepherd would not break, no matter how the sea chose to speak.

From the cliffs above, a single window burned with lamplight. Behind its glass, Dr. Olivia Turner stood frozen, ink-stained hands pressed to the sill. Her notes lay scattered across the table, pages curled by the salt air, sketches of bell-shaped bodies and tendril arrays stark against the fluttering shadows. The careful order of her work, measurements, ratios, Latin names, looked suddenly childish in the face of what darkened the sky.

Her eyes tracked the swarm as it drifted higher, blotting out the stars one by one until even the constellations seemed to retreat. She had dreamed of unlocking nature's secrets, of giving the sea resilience against a world that had bled it dry. She had chased answers in petri dishes and tanks, whispering promises into glass like prayers. But nothing, no experiment, no fevered midnight calculation, had prepared her for this.

These were no longer curiosities under her care, no longer the specimens she told herself she was improving. They were creatures freed from the sea itself, pulsing with a rhythm that seemed older than stone, older than tide. In their glow, the very laws of her world bent like reeds before a gale.

Olivia's breath fogged the glass. For one fractured instant, she saw her own reflection superimposed over the swarm, her eyes lit with the same unearthly sheen, her ink-stained fingers ghosted by trembling tendrils of blue. She staggered back from the sill, clutching her burning wrist through its wrappings, the glow beneath her bandage answering the sky above.

"God help me," she whispered, though she no longer knew if it was a plea, a confession, or a prayer.

Below, the church bell gave one final, strangled toll before the rope snapped, its bronze tongue caught mid-swing. The broken peal wavered in the air for a breath, then was devoured whole by the drone.

The hum pressed down from every side. It didn't just echo in ears, it ran through rafters, seeped into the cobbles, crawled up marrow, until every chest in Haven's Reach seemed to beat in borrowed rhythm. Windows rattled in their frames. Nails shrieked as boards strained. Dogs whimpered beneath porches, muzzles buried against dirt as though the ground itself might shield them.

People clutched each other in the darkened square, faces turned upward, eyes squeezed shut; it no longer mattered where the sound came from. It was everywhere. The sea. The sky. The stones.

The very bones of the town vibrated with it, as though Haven's Reach itself had become an instrument strung to the swarm's will.

And from her window on the cliffs, Olivia knew, this was no storm, no accident of nature. This was a summons.

Olivia's reflection in the window flickered in the jellyfish light, her face broken into shards of pale green and cold blue. For the first time, she saw fear there, not the quiet dread she had carried since her experiments began, but something sharper, heavier. Fear not just for herself, but for Haven's Reach. For the children pressed beneath their mothers' shawls. For the fishermen whose nets had been reduced to ash in their hands. For the town that had never once imagined the sea might turn against them.

The swarm pulsed again. Light rippled over rooftops and cobbles, washing the town in a cadence too measured to be chance. The hum deepened, sinking like an anchor into every heart, and the people below cowered as one body, as though the very air had become a net cast over them.

In that moment, Olivia understood what the townsfolk could not name: the creatures weren't merely drifting, nor were they simply hunting. They were watching. Measuring. Waiting for an answer.

Her bandaged wrist burned hot against the glass, the faint glow beneath the linen throbbing in perfect time with the swarm's terrible heartbeat.

DISTURBANCE

T he hum rose in pitch until it rattled teeth, a vibration that seemed to crawl through marrow and shake the nails from the boardwalk. The dogs, which only moments earlier had barked in furious defiance, grew quiet one by one. Their growls turned to uneasy whines; tails pressed between their legs as they shrank against their masters' boots.

Only one remained, an old hound, broad-shouldered and stubborn, the same that had broken loose earlier. It stood rigid at the water's edge, hackles raised high, its eyes fixed on the drifting lights above. The sound that rumbled from its chest was not a bark now but a low, unsteady growl, more fear than fury.

Then a tendril, slender, luminous, and impossibly long, descended from the swarm. It did not strike but brushed across the animal's flank with the lightest of touches, like a ribbon trailing in a breeze.

The hound stiffened, a sharp yelp tearing free as its body jolted in a sudden spasm. Its paws scrabbled against the planks, claws clattering in a frantic rhythm, and then it turned, bolting down the lane at a dead run. Its form, once solid in the lamplight, dissolved into the glow, swallowed by shadows before the crowd could blink.

It did not return.

A ripple of breath passed through the villagers, as though the whole wharf exhaled at once. A mother pulled her child hard against her chest, whispering a prayer into his hair. Her lips trembled on the words, half-spoken and half-silent. A fisherman crossed himself, his eyes darting skyward. Others said nothing at all, too afraid to give voice to what they had just witnessed.

The hum pressed on, unbroken, filling the silence left in the dog's absence. The creatures drifted higher, their tendrils trailing like pale banners, as if the hound had never mattered at all.

Then came the fishermen. Three men wrestled to haul their small boat ashore, its nets still dragging heavy with the weight of their catch. The hull scraped across the sand in a harsh groan, water spilling in rivulets that gleamed faintly under the shifting glow above.

They heaved together, shoulders straining, but the boat refused to come fully free of the tide. The sea seemed to grip it; each pull stronger than the last, as though unseen hands beneath the surface were claiming what was never theirs.

A gasp rippled through the onlookers as one of the creatures drifted lower, descending like a pale lantern let loose from the heavens. Its tendrils trailed downward, long and fluid, ribbons of fire and glass that swayed as if tasting the air. With a sound like a sigh, one tendril brushed the side of the boat.

Where it touched, the wood glowed faintly, spreading outward in ripples of sickly light. The surface blackened, not with flame, but as though the very life had been drawn out of it, leaving only a hollow shell that still held its shape.

"Cut it loose!" one of the fishermen shouted, panic sharpening his voice until it cracked. He fumbled for his knife and slashed wildly at the nets, but the fibers glowed beneath his blade, curling tighter as if resisting him. The tendrils slithered around the ropes with deliberate grace, and with one final pull, the catch was gone.

The water flared with eerie radiance as the fish writhed, their scales reflecting the glow in unnatural flashes. Silver bellies rolled once, twice, before sinking, their light fading into the black.

The man nearest the net stumbled back with a strangled cry. His sleeve had been grazed by a trailing tendril. Angry welts bloomed across his forearm, glowing faintly as though the swarm had left its mark beneath his skin. He clutched at the wound, his knees buckling in the sand, breath tearing ragged from his lungs.

A sour tang filled the air, sharp as iron and salt, carried on the wind. The crowd recoiled, hands pressed to mouths, their fear thickening with each pulse of the hum.

Panic followed. The boardwalk erupted in a frenzy of shouts and scraping boots as townsfolk shoved past one another, desperate to reach the safety of their homes. Mothers gathered children into their arms, skirts whipping around their ankles as they ran. Old men stumbled, caught in the press of bodies, their voices lost in the din.

Lanterns guttered and flickered, their flames shrinking as though starved for air. One by one, they winked out entirely, as if the creatures themselves were drinking the light straight from the wicks. Shadows surged in their absence, broken only by the eerie shimmer cascading from above.

Doors slammed. Shutters clapped closed so hard the wood splintered. In the space of a few breaths, the lively boardwalk lay empty, abandoned to the glow and the hum.

But the safety of walls offered no true refuge. The light seeped through every crack and seam, painting ceilings and floors with ghostly ribbons of green and blue. Inside, the hum grew worse, vibrating through beams, crawling up plaster, shaking crockery loose from shelves until plates and mugs shattered across stone floors.

Babies screamed in their cradles, their wails merging with the drone until it was impossible to tell one from the other. Windows rattled in their frames, straining against unseen pressure. A horse, panicked in its stall, kicked hard enough to splinter wood.

Then, sharp and startling, came the crack of a rifle. Somewhere in the confusion, a trembling hand had misfired. The shot tore through the air and vanished into the drone, devoured before its echo could even return.

The hum rolled on, undisturbed.

On the edge of town, Mrs. Callahan collapsed to her knees, her shawl slipping from her shoulders as though the strength had gone out of her bones. She rocked back and forth, eyes wide and glassy, lips trembling with words only half-formed.

"Not of this world… not of this world…" she muttered, her voice cracking under the weight of terror.

Her rosary slipped from her fingers, tumbling into the sand. The beads scattered in a quick cascade, clicking softly as they rolled until the strand broke apart entirely. Under the shimmer of the jellyfish glow, they shone like scattered seeds, small orbs of faith spilled uselessly across the earth.

She clawed at them with shaking hands, trying to gather the beads back, but each time her fingers closed, another slipped through. The drone pressed harder, rattling in her chest until her prayers dissolved into sobs.

Children passing with their parents stared, their eyes wide. One whispered, "She's praying the wrong way." Another was pushed and pulled along, but the words lingered. To them, Mrs. Callahan's collapse was not just fear but proof: even the old woman who'd watched the tides all her life knew this light was no storm, no sky-born wonder. It was something else. Something that didn't belong.

Children buried their faces in their mothers' skirts, their sobs muffled against coarse wool as the women clutched them tight, whispering prayers that fell flat beneath the drone. The men stood in ragged knots along the boardwalk; rifles clutched in white-knuckled grips.

But courage faltered in the glow. Each time a tendril drifted lower, swaying like some pale whip of light, the barrels of their guns wavered and sank. The rifles sagged toward the ground, the men's eyes wide and hollow. They had faced storms and starvation, shipwrecks and famine, but this, this was not a foe made of flesh or wood or wave. The creatures floated too high, too alien, too untouchable. No weak point to strike, no anchor to cut, no weight to drag them down. Bullets felt like pebbles thrown at the tide, and the men knew it.

One man raised his rifle anyway, teeth bared in defiance, but his neighbor caught his arm and hissed through clenched teeth, "You'll only anger it." The two stood locked for a heartbeat before the would-be shooter lowered his weapon, his hands trembling as though the rifle itself had turned against him.

Above, the jellyfish pulsed again, their glow cascading across the rooftops, and Haven's Reach seemed to shrink beneath them, its defenders small, and unarmed in all the ways that mattered.

A silence heavier than thunder followed, broken only by the steady pulse of light overhead. Every eye turned skyward. The swarm hung there, suspended in their eerie rhythm, as though waiting for some unseen signal.

And high on the cliffs, a lone figure watched. Olivia Turner stood with her lantern shielded, its flame reduced to a pinprick, her notebook braced against the wind. Each pulse of the jellyfish reflected in the lenses of her glasses, ghostly light etched across her face in fleeting patterns. Her hand moved ceaselessly, ink scratching across paper: rough sketches of tendrils, numbers scrawled in quick, cramped columns, and half-formed questions in the margins.

She worked as though the act of recording could keep her steady, as though if she named the shapes and measured the rhythm, she could strip the terror from them. But the hum slid under her skin, turning her pulse into its metronome. Her lips moved unconsciously, counting, calculating, already seeking a logic she feared she might find.

Her pen faltered only once. Below, the fisherman's cry tore across the water, raw, pained, and terribly human. Olivia's breath hitched, the sound striking her chest like a blow. For the first time that night, she stopped writing. The hum filled the pause, louder in the absence of her scratching.

She shut the notebook slowly, the faintest shadow passing over her face. Her fingers tightened on its cover; knuckles pale against the dark leather.

"It's begun," she whispered. The words barely left her lips, but the hum shifted as if in answer, carrying her voice into the night.

For one disquieting heartbeat, it felt as though the creatures were listening.

WORLD

The cliffs were the only place Olivia Turner felt at home. Each step up the narrow path carried her further from the murmur of the harbor and deeper into the rhythm of sea and stone. Wind scoured the rocks, tangling her dark hair and tugging at her coat, sharp with salt. Spray leapt high with every crashing wave, turning the climb treacherous even in daylight, slabs slick beneath her boots, the air damp with brine. Gulls wheeled overhead in noisy squadrons, their cries darting through the twilight like accusations.

Below, Haven's Reach clung stubbornly to its crescent of harbor, roofs hunched low against the sea wind.

From this height, the town looked fragile, almost temporary, a scattering of shells that the tide might sweep away at any moment. Olivia paused, breath clouding in the dusk, and for a fleeting second, she felt the weight of both distance and belonging. She knew these people, their laughter and their suspicions, their prayers and their rumors. Yet she had chosen the cliff instead, the hard solitude, the endless horizon, the sanctuary carved halfway into stone where her work waited like a promise she couldn't abandon.

Ahead, her laboratory rose from the cliffside like an extension of the rock itself, a glasshouse set into weathered stone. In the fading light, its tall windows gleamed with reflections of the sea below, throwing back ghostly fragments of wave and sky. To the villagers, it was a curiosity at best, an omen at worst. To Olivia, it was no eccentric beacon; it was sanctuary.

From that height, Olivia always felt suspended between two worlds: the human one below, with its narrow minds and whispered judgments, and the ocean beyond, endless and alive.

Her laboratory jutted from the cliff face like a lantern, half glass, half stone. By night, the townsfolk called it the glasshouse, claiming its tall windows glowed unnaturally against the sea. Sailors swore they could glimpse its light from ships miles offshore, mistaking it for a beacon or some wayward star. To Olivia, it was no eccentric beacon; it was sanctuary.

It had not always been hers. As a girl, she had danced in these tide pools, splashing with bare feet while her mother's laughter carried across the rocks. Her mother had once dreamed aloud of a house here, built into the cliff, walls of glass that would hold both sky and sea in their reflection. Olivia had never forgotten the way her mother's eyes had shone with that imagining, as though she could already see herself living there, balanced between land and water.

But then her family had sent Olivia inland, away from the sea, away from everything she loved. They'd called it schooling, opportunity, a proper education. She had obeyed, but every season that passed away from the tides felt like an exile. By the time she returned, the dreamer was gone. Her mother's laughter had vanished with her, her father buried in the churchyard not long after. Only her grandfather remained, stern, weathered, and unwilling to leave the cliffs that had claimed so much.

So, Olivia built the house her mother had once imagined, not for grandeur, but for memory. Its glass walls were less a laboratory and more a shrine, a living echo of the dream they had shared in her childhood. Here, when the wind howled through the gullies and the surf struck sparks against the rock, she could almost hear her mother's voice again, urging her to dance barefoot in the foam, reminding her that the sea was not just hunger and danger, but life.

Inside, the air was warm and damp, heavy with the briny tang of seawater laced with the faint metallic bite of iron. It smelled like the tide pools of her childhood, distilled and contained within stone and glass. A constant undertone filled the room, not silence, never silence, but a mechanical hymn: the steady thrum of pumps, the hiss of heaters, the rhythmic pulse of oxygen regulators, and the burble of water circulating through pipes. It was a symphony of control, human-made currents mimicking the ocean's heartbeat.

Rows of aquaria lined the walls, their panes gleaming under chains of hanging lamps. Some were small, fit for the palm of her hand; others rose taller than her shoulders, their glass curved to hold whole tides in miniature. Within them, jellyfish drifted like fragments of fallen stars, their bells pulsing with ghostly grace, tendrils unfurling like ribbons of spun glass. Their soft light painted the walls in hues of violet, green, and blue, each pulse a reminder of the fragile balance she tried so desperately to command.

Olivia exhaled, tension leaving her shoulders. Here, she belonged. The town below might spit the word Turner like a curse, but within these walls, her name and her work finally carried weight.

She set down her satchel on the oak workbench and moved seamlessly into her routine, every motion an old ritual. Notebook open, pencil sharpened to a fine point, she began her nightly circuit. Each gauge was read aloud under her breath, each dial adjusted by hand: thermometers checked for fractions of a degree, salinity meters studied for subtle shifts, oxygen flow regulators nudged until their needles aligned.

Every number mattered. Every fluctuation could mean the difference between success and collapse. To the fishermen, the ocean was prey and peril. To Olivia, it was pattern, rhythm, and potential, all things that could be measured, altered, improved.

And tonight, more than ever, she needed to believe control was still possible.

"Specimen Twelve," she murmured, tapping lightly against the glass of a small tank where a pale jelly drifted in lazy spirals. Its bell contracted once, twice, then flared with a soft glow before dimming again. "Response time improving. Good. Very good."

Her pencil hovered a moment before scratching across the page in brisk, precise lines. Her handwriting was neat when she wanted it to be, but tonight it wavered, the edge of her hand smudging the wet ink. She forced the notes into order, but beneath them, smaller script, hidden from any eyes but hers, she wrote a truth she would never admit aloud: *If this works, it won't just be survival. It will be renewal.*

She paused, the words stark on the page, then moved to the next tank.

This one was larger, its glass walls braced with iron ribs that hummed faintly under the pressure. Within it, three jellies circled in deliberate formation, their tendrils trailing like banners in unseen currents. She had altered their genomes only weeks before, splicing strands together with the precision of a craftsman shaping destiny. The intention had been simple: resilience. Creatures that could endure what reefs and fish could not, waters growing warmer, darker, emptier with each year.

But these were not simple.

Their tendrils stretched longer than her notes predicted, brushing the glass in delicate, almost searching strokes. Each time they touched, the faintest shimmer spread across their bells, as though they were tasting the limits of their prison.

"Not too much," Olivia whispered, fingers tight on the nutrient valve as she adjusted the flow. Her breath fogged faintly against the tank. "Grow, but not uncontrolled. Controlled, always controlled."

She lingered longer than she meant to, eyes tracing the pulse of their light, the strange synchrony of their movements. Something about it prickled at the back of her mind. They were changing faster than the others, too fast, perhaps. Faster than the models, faster than her calculations allowed.

Her grandfather's voice surfaced in memory: *The sea does not bend to rules written on paper, child. It has its own.*

Olivia pressed her lips together and forced the thought away. Risk was the price of progress. The reefs below were dying, the ocean emptying with each season. To hesitate now was to lose everything.

Still, she could not help the flicker of unease that lingered as she wrote her final note in the margin:

Unexpected acceleration. Monitor closely.

The words looked fragile on the page, as though even the ink feared to hold them.

Her gaze flicked to the far corner of the lab, where a battered wooden crate rested beneath a draped cloth. The sight of it tugged at her more than the glowing tanks ever did. She hesitated, then crossed the room and lifted the cloth with careful fingers.

Inside lay the failures.

Rows of jars sat sealed in saline, glass fogged with salt crystals. Each one contained a reminder of where ambition had overstepped its bounds: bells collapsed inward like shriveled fruit, tendrils frayed into pale strands, bodies warped into grotesque knots of translucent tissue. She rarely looked at them for long. Sketches in her notebook carried the memory well enough, jagged outlines she had drawn with a trembling hand on the nights she couldn't sleep.

And yet she could never bring herself to destroy them.

They stayed here, shadows bottled in glass, silent witnesses to the thin line she walked between discovery and catastrophe. Every time she considered discarding them, something stopped her, a sense of duty, perhaps. Or guilt.

But it was the last jar that haunted her most, the one that was no longer here.

Just weeks ago, she had carried a specimen down the cliff path, its form weak, its bell collapsed into a sickly curl. It had looked ready to dissolve into nothing. A mercy, she told herself then. Better the open water than slow decay in glass. She had waded into the tide pool until seawater lapped at her ankles and tipped the jar into the swell.

For a moment, the creature had drifted limp and lifeless. Then, as the waves took it, something impossible happened.

It glowed. A soft, startling blue, faint but undeniable, spread through its bell as though the ocean itself had woken it. Olivia froze, heart hammering, watching as the light brightened once, twice, before the tide carried the fragile body into deeper water.

No one else had seen it. No one else ever would. But the memory lingered, sharp as salt on her tongue. Had she imagined it? A trick of the light, a moment where desperation turned seawater into a miracle?

She pulled the cloth back over the crate with quick, restless hands, but the image of that weak pulse of blue followed her, settling deep in her chest.

For the first time that evening, she felt less like the keeper of her glasshouse and more like a guest in it, watched, measured, and quietly judged by the sea just beyond her walls.

The town thought her reckless. Some called her "the hermit on the cliffs." Others whispered that her experiments disturbed the tides, stirring things that were better left to silence. Olivia didn't care. What was a little suspicion compared to the chance to preserve the only world she had ever truly loved?

And yet, she wasn't deaf to it. On certain nights, when the wind carried voices from the harbor up the cliffside, she caught the sharp edge of her own name wrapped in words like *unnatural* or *cursed*. They thought her meddling with

God's work. She thought of it as listening to what the ocean had to say.

Her grandfather heard the same whispers, though never with surprise. As one of Haven's Reach's elders, his word carried weight among fishermen and townsfolk alike. He spent many evenings at Olivia's glasshouse, easing himself into the old armchair by the hearth. Though a guest room stood ready, he rarely took it. Instead, he chose the living room, his presence a quiet, visible bridge between his granddaughter's "fancy instruments" and the village's old ways. He spoke to them in the language of respect and balance, reminding them that Olivia meant no harm, that her hands were guided by the same reverence for the sea they all claimed to hold.

Not everyone believed him, but few dared to challenge him. His voice, even softened with age, could cut through rumor like a tide against stone.

And then there was Alex Shepherd.

He had been part of her grandfather's shadow since Olivia was fifteen, when she was preparing to leave for school inland. An orphan by sixteen, Alex had found no family but the sea and the old man who took him under his wing. He followed her grandfather everywhere in those years, learning the currents, the knots, and the patient reading of weather that meant survival. Broad-shouldered and strong even then, Alex became his steady hand when the elder faltered, his quiet strength a kind of anchor.

Alex never joined in the gossip about Olivia. He never repeated the words others used for her, never called her *hermit* or *cursed*. Instead, when tavern talk grew too sharp, he silenced it without raising his voice. A single look, a word spoken low, and the conversation shifted. "You don't understand what she's doing," he'd say simply, and that was enough. No one wanted to test Alex's patience, he carried the kind of authority that didn't need to shout.

without flaunting it.

Olivia had known him only at a distance then, catching glimpses of him on the docks when she returned from lessons, tall and taciturn at her grandfather's side. He had always seemed older than his years, hardened by loss and salt wind. She never imagined back then that the boy who trailed her grandfather's steps would one day stand quietly between her and the weight of Haven's Reach's judgment.

She pressed her palm to the glass of the largest tank, watching the jellyfish respond with a slow, luminous pulse. The glow painted her skin in soft blue light, making her look, for an instant, less like a scientist and more like a creature of the deep herself.

Her breath slowed to match their rhythm, in, out. Pulse, fade. The lab's hum receded until all she heard was the silent language of their movement. She closed her eyes and let it wash over her, imagining herself weightless, drifting in the currents as they did. For a fragile moment, she felt as though she had crossed some invisible threshold, no longer an observer, but part of their world.

Her mind slipped back to her mother's voice, teaching her the old songs that mimicked the tides, the dances in the tide pools when she was small. *The ocean remembers, Liv,* her mother had said. *It only waits for us to listen.* The memory ached now, sharper than the salt air.

She opened her eyes again. The jellyfish pulsed brighter in reply, as though they too remembered. A soft ache stirred in her chest, half hope, half dread.

In that silence, she almost believed the ocean could be saved, that her hands, steady on the glass, might shape the future instead of destroying it. That her mother's dream, her own lifelong obsession, could still be more than a fragile illusion.

But then, the faintest vibration shivered through the tank, barely perceptible yet enough to lift the hairs on her arms. It faded quickly, gone as if imagined. Olivia blinked, heart stuttering. The jellyfish drifted on, their pulses steady, innocent. Still, the sensation lingered.

Not for the first time, she wondered if the ocean was listening back.

Four

FIRST HINTS OF CHANGE

Olivia bent over the middle tank; pencil poised above her notes. The three jellies she had altered last month no longer drifted in lazy spirals. Their movements carried intention now, as if some invisible conductor guided them. Their bells pulsed in perfect rhythm, one, two, three, tendrils brushing the glass with a soundless insistence, as though they sought something beyond the confines of water and light.

Her brow furrowed. She leaned closer, watching every detail, her breath fogging the glass.

"Synchrony… unrecorded in previous specimens," she murmured, the words clipped and hushed. Her pencil scratched furiously across the paper, her script tight and hurried. "Duration: ninety-three seconds and counting."

The stopwatch in her other hand ticked with cruel precision, each click intensifying her pulse. Ninety-four.

Ninety-five. The glow of the jellyfish reflected off the polished brass face, making the seconds gleam cold and sharp. By the time the creatures finally broke formation, scattering like leaves in a current, Olivia's palms were slick, her hand trembling as she set the watch down.

They had never behaved like this before.

The scientist in her burned with curiosity, a gnawing hunger she couldn't ignore. She adjusted the oxygen feed, twisting the valve with careful fingers, then dimmed the overhead light until the lab was cast in shadow. The rows of tanks glowed brighter in the gloom, each one a beacon of cold fire. Shadows of the jellies' bells rippled along the walls, strange and distorted, as if the room itself had sunk beneath the waves.

Within moments, the altered trio regrouped. Their bells contracted in flawless symmetry, each pulse resonating with the next. They swam forward, not wandering but aligned, their tendrils pressing flat against the glass in unison. The sight tugged at something deep in her, fascination, yes, but also a thin strand of unease.

It looked less like animals adapting to their environment and more like a choir finding its voice.

Olivia leaned in until her forehead nearly brushed the cool glass. The faint vibration hummed against her skin, so subtle she might have imagined it, but her instincts whispered otherwise. Her notes lay forgotten, her pencil hovering uselessly in her grip as the thought settled, heavy and inescapable:

They weren't just moving.

They were communicating.

The hum came then. At first, it was so faint she almost dismissed it, a sigh through the vents, the whisper of wind pressing against the cliff face. But as she stilled, breath fogging the glass, she realized it wasn't in the air around her at all. It was inside her. A low vibration thrummed through her chest, a resonance that seemed to rise from the very stone beneath the lab.

The familiar whir of pumps, heaters, and bubbling regulators dimmed, fading into an almost sacred silence.

As though the machinery itself bowed under the weight of the new sound, the jars of pencils on her desk quivered, their tips knocking together in a faint, nervous rattle. On the shelf behind her, a beaker of saline rippled, the surface shimmering with concentric rings that spread outward in perfect rhythm with the pulse of the jellyfish.

Her bones vibrated with it. For a heartbeat, her pulse fell into step with the sound, her body betraying her mind syncing as though it had always belonged to this strange new rhythm. Then the resonance surged past, racing faster, deeper, until her chest ached with the effort of resisting it.

Olivia staggered back from the tank; one hand braced against the workbench. Her heart hammered, not only with fear but with awe.

"No," she whispered, the word cracking in her throat. "Impossible."

Her training demanded discipline. She forced her trembling hand to the page, scratching hurried words into her notebook even as her script wavered beneath the tremor of her fingers:

Audible vibration, possibly subsonic frequency emitted by altered specimens.

The tip of the pencil slipped, skidding across the page. She froze, eyes flicking upward.

Inside the tank, the glow surged, flaring brighter for the span of a breath. Not random. Not coincidental. It pulsed in time with her hesitation, as though the creatures had answered her lapse in certainty with one of their own.

Olivia's throat tightened. Her hand hovered over the page, unwilling to commit the word that pressed at the edge of her thoughts.

Aware.

The room seemed smaller suddenly, the air pressing in, thick and humid, as though the walls themselves exhaled with the tanks. The tang of iodine clung sharp at the back of her throat, mixing with the metallic edge of salt and iron. Droplets of condensation trickled down the glass panes, catching the eerie light and fracturing it into ghostly shards across the floor.

Olivia pressed her fingertips to her temples, kneading slow circles into her skin. The gesture was meant to soothe, but her hand still trembled faintly. She thought of the months, years, of trial after trial, petri dishes abandoned, samples discarded, notes scrawled and struck through until her margins were nothing but failures. All those long hours bent over microscopes, chasing something she could never quite reach.

She couldn't allow herself to break now. Not when she was closer than she had ever been. Not when she could feel the answers, no, the changes, stirring on the other side of that glass.

Her eyes flicked back to the tank. The jellyfish hovered there, bells expanding and contracting in uncanny unison, tendrils stretched like fingers seeking purchase. Their glow rippled faintly, as though waiting for her acknowledgment.

"It's only data," she whispered, forcing the words through the knot in her chest. "That's all it ever is."

But the sound of her own voice rang hollow in the charged air, like a lie too brittle to bear its weight.

When she reached out to adjust the salinity gauge, one of the jellyfish drifted forward, slow as a shadow. Its bell pressed against the glass exactly where her hand rested, tendrils unfurling like filaments of pale fire. They spread outward until they framed the outline of her palm, glowing softly, as though the creature were mirroring her.

Olivia froze, breath locked sharp in her chest, ribs aching with the effort of stillness. The hum deepened, low and invasive, pressing into her eardrums until her vision blurred at the edges. Her heart hammered, but even that rhythm seemed stolen, overtaken by the steady pulse radiating from the tank.

For one terrifying instant, she felt stripped bare, as though the creature could see not her skin or her bones, but the core of her, memories, guilt, secrets she hadn't admitted even to herself. It wasn't a gaze. It was recognition. A vast, alien mind brushing against hers, impossibly old, impossibly patient.

Then, abruptly, the moment shattered. The jellyfish released the glass, drifting back into the tank as though the contact had never happened. The others followed, pulsing in aimless patterns, their glow soft and deceptively gentle. The hum receded, leaving only the ordinary hiss of pumps and the ticking clock on the wall.

Olivia staggered back a step, dragging her hand from the glass. Her laugh came thin and brittle, too sharp in the still room. "Coincidence," she said aloud, the word as fragile as spun sugar. "That's all."

But when she bent over her notebook again, her hands betrayed her. The precise, clean lines she prided herself on fractured across the page. Her notes bled into crooked scrawls, letters smudged and uneven, as if the pencil belonged to someone else.

Her pulse throbbed erratically beneath her skin. She pressed trembling fingers against her wrist, trying to steady it, but the beat seemed wrong—out of sync, echoing in a strange counter-rhythm to the fading thrum of the tank. Sweat rolled cold down her temple, stinging her eyes.

She shut the notebook with a snap, but her gaze refused to obey. It kept straying back to the glass, to the drifting bells and slow tendrils. They looked harmless now, delicate, ethereal, nothing more than creatures carried on invisible currents.

Harmless. That was the word she repeated in her mind, like a prayer.

And yet, every time she blinked, she still saw her palm imprinted in ghostly light against the tank wall, as if some part of her had been marked.

Five

ESCALATION IN THE LAB

The pumps hissed and burbled in their usual mechanical rhythm, but today that rhythm was swallowed by something deeper.

The hum. It emanated not just from the water but from the very frame of the tank, a vibration that turned every note of machinery into a haunting accompaniment, as though the room itself had been tuned to a hidden frequency.

Her temples throbbed. She rubbed at her eyes, blaming the fatigue that fogged her vision. But when she looked again, the glow inside the glass flared brighter, each pulse timed so perfectly that it seemed less like coincidence and more like will.

She set down her notebook, unease prickling at the back of her neck like static. The lab should have been cool at this hour, the stone walls usually keeping the dawn chill trapped inside. Yet today the air hung heavier, warmer, carrying with it a faint metallic tang, sharp as rusted iron on her tongue. It clung to her throat with every breath. A bead of condensation gathered on the inside of the tall window, sliding down the pane before vanishing into the sill. Outside, frost silvered the cliffs, but within her sanctuary, the atmosphere felt almost fevered.

"Specimen group twelve," she said at last, forcing her voice into the same clipped tone she had always used for observation. Naming them, cataloguing them, it had always been a shield, the ritual that kept her tethered to control. "Activity level increased by thirty percent overnight."

The words echoed faintly off the stone, but what unsettled her was how the tank seemed to echo them too. The nearest jelly shifted with uncanny precision, drifting forward until its translucent bell hovered against the glass. It contracted once, twice, the rhythm eerily matched to her syllables, as though the creature were listening.

Olivia's pulse leapt. She swallowed hard, but her pen scratched down the margin anyway: "Possible correlation between vocalization and response. Further testing required."

She took a careful step closer, the wooden floor groaning beneath her weight. At once, the hum sharpened. What had been a low drone became something finer, thinner, resonant enough to vibrate against her ribs, as though her bones had become part of its instrument. Each word she spoke didn't just leave her lips, it seemed to be caught and amplified, folded back into her body by the unseen resonance.

Her lips pressed together. The jelly pulsed again, a beat behind her heartbeat. Deliberate. Waiting.

Her breath hitched, sharp in the back of her throat. Every instinct screamed at her to step away, but she forced her trembling hands forward, closing them around the cold metal of the nutrient valve. Stay steady. Stay in control. If she broke routine now, the whole system would unravel.

The nearest jelly convulsed. In a sudden burst of motion, its bell contracted and flung it forward, tendrils snapping against the glass with the speed of whips. The impact rang through the lab like a struck bell. The entire tank jolted against its stand, sending water splashing hard against the lid.

A jagged crack lanced across the pane, thin and merciless as a bolt of lightning. It hissed as it spread, branching outward in quicksilver veins that glowed faintly in the jellyfish light. The sound was sharper than nails dragged across slate, and it went straight into her bones.

"Damn!" Olivia staggered back, nearly tripping over the stool behind her. Her heart slammed against her ribs as though it were trying to escape.

She lunged for the rag at the edge of the workbench, pressing it desperately against the fracture. The fabric darkened instantly, soaked through as water welled in cold beads, dripping down the glass in steady rivulets. Each droplet struck the floorboards with a tick so regular that it sounded like a clock counting down.

The crack deepened under her hands with another ominous hiss, the pane bowing ever so slightly, as if the pressure inside was testing its prison. Her arms strained against the trembling tank, the rag slipping in her wet fingers. She bit down on her lip until she tasted copper, whispering to herself, "Hold. Just hold."

The jelly pressed forward again, its bell flattening against the glass as though it could sense her weakness through the barrier. Its glow brightened, saturating the lab until the crack shimmered with a heatless brilliance, veins of blinding light crawling outward with every pulse.

Then, so subtle at first she thought her eyes deceived her, a filament slipped through. A single tendril, impossibly thin yet rigid as drawn wire, forced itself out of the fracture.

It moved with dreadful patience, writhing in the open air like something tasting freedom.

Olivia froze, breath caught. Her instincts screamed at her to pull back, but her body was a fraction too slow.

The tendril struck.

It whipped across her wrist with a speed that made her flinch as if struck by lightning. The contact seared, not like fire on flesh but like liquid flame poured directly into her veins. She cried out, the sound ripped raw from her throat, and the rag fell uselessly from her grasp.

Her hand clamped reflexively over the wound, but the gesture offered no comfort. The pain roared beneath her skin, blooming outward in violent waves. She gasped as blue light erupted beneath the surface, faint at first, then flaring brighter, racing in branching lines along her veins. They climbed her arm like ivy made of fire, each pulse throbbing in perfect synchrony with the jellyfish still pulsing behind the glass.

Her knees buckled. The world tilted. She caught herself against the edge of the workbench, knuckles whitening as she fought to stay upright. The hum surged with her pain, resonating through her ribs and skull until it seemed she had become part of it, a living note in their unholy chorus.

For a heartbeat, she swore her own pulse had been overwritten. Each throb in her chest matched theirs exactly, her body answering the rhythm of the tank as if she were no longer entirely her own.

And then, as suddenly as it began, the jelly recoiled. Its glow dimmed to a softer radiance, the tendril sliding back through the crack as though sated. The hum subsided into a low, steady vibration that lingered in her chest like an echo refusing to fade.

Olivia sagged against the workbench, trembling. Her wrist burned, the veins still flickering with a faint blue light that slowly dulled into a sickly shimmer. She pressed her hand to the wound again, more in desperation than belief, as if she could trap whatever had entered her beneath the surface. But the rhythm in her blood told her it was already too late.

Olivia collapsed onto the stool, every breath a ragged draw that scraped her throat. Her wrist throbbed in waves, each pulse a flare of molten ache that made her jaw lock tight. She clutched the pencil with fingers that refused to obey, the tremor running so deep she feared she'd snap the graphite before she even touched the page.

Her teeth ground together as she forced her other hand, her uninjured one, over the notebook, dragging the pencil across the paper. The words came jagged, slanted, breaking from their usual precise order. Her neat script fractured into uneven scratches, letters half-formed and smeared where her wrist brushed the page.

Specimen breach. Subject sustained contact burn. Immediate reaction: radiating pain, luminescence under skin, subsiding within thirty seconds.

She read it back, breath hitching. The statement looked clinical, detached, exactly the way she wanted it. Yet the smudges and shakiness betrayed her; the page itself trembled with the shake in her hand. Ink pooled where the pencil had paused too long, tiny blotches that felt like confessions she couldn't scrub out.

Olivia set the pencil down with a clatter, her arm limp at her side. The glow beneath her skin pulsed faintly still, each line of light a quiet reminder that she had crossed a boundary she couldn't redraw. She pressed her palm hard against the band of her wrist, as though pressure alone might erase what now lived there.

Forcing her gaze upward, she met the tank's glow. The jellyfish hovered calmly once more, as if they had never lunged, never breached, never touched her. Their bells rose and fell with perfect serenity, tendrils drifting in lazy ribbons. Innocent. Harmless.

Her laugh cracked in her throat, brittle and thin. The words she had written sat stark on the page, but the true entry she couldn't bring herself to write was simple: *They know me now.*

Her laugh faded as quickly as it came. She stared down at the notebook, at the words she had forced onto the page, but the longer she looked, the less they resembled truth. Data points and measured terms could not contain what had just happened.

The ache in her wrist pulsed again, this time in eerie harmony with the glow inside the tank. She didn't need to check her pulse to know it was out of step with her own body, her heart raced, frantic and erratic, but the light beneath her skin beat with calm precision, echoing the rhythm of the jellyfish.

She dragged the band of her sleeve down over the welt, hiding it from view, though no one was there to see. For a long moment, she kept her palm pressed hard against it, as if secrecy alone could undo it.

Her gaze lifted back to the tank, to the slow rise and fall of translucent bells, the lazy drift of tendrils curling like smoke. Innocent. Harmless.

That was what she told herself.

But in the back of her mind, louder than the hum, louder even than the racing of her heart, a single thought whispered like a truth too terrible to name:

They know me.

The lantern flickered once, then guttered, and steadied. Olivia sat frozen in the dim light, notebook open before her, the last word she had written smeared by her trembling hand.

The room felt colder now, and for the first time, she wondered if her sanctuary had become a cage.

Later that week, rumors spread through Haven's Reach like sparks skittering over dry grass.

By the docks, fishermen bent over their nets, their voices low but sharp. They whispered of strange lights shining from the cliffs at night, too bright, too steady to be lanterns. Some swore the glow reached down into the tide pools, staining them blue, so that the water itself seemed to breathe with Olivia Turner's touch.

Others brought back nets heavy not with fish but with scraps, translucent ribbons that writhed faintly in the moonlight before dissolving into nothing but brine. Men spat into the sea when they spoke of it, muttering that the catch had thinned since she'd started her work.

Children turned the fear into dares, challenging one another to climb the path to "the witch's glasshouse." They would creep no further than the first switchback before the faintest hum floated down on the salt wind. Their bravado shattered; they would scatter back to the square, shrieking with laughter that rang hollow in their throats.

At the market, wives haggled with their purses clutched tight, whispering behind shawls. "The gulls don't roost by the cliffs anymore," one said, eyes darting. "Birds know better."

Another crossed herself, murmuring that the fish tasted strange, though no one dared admit aloud that they still ate them.

Every small oddity found a tongue, and every tongue found an audience. A candle that sputtered in the night became a sign of her curse. A child's fever became her fault. Whispers grew teeth, and teeth turned into blame. And though no one spoke Olivia's name too loudly, every glance that turned toward the cliffs carried its weight.

Her grandfather appeared one evening, lantern in hand, his face creased with worry. The flame inside sputtered each time the hum trembled faintly through the walls, as though even fire feared to burn too brightly in the glasshouse.

"Olivia," he said, standing in the doorway of her lab. His shadow stretched long across the wet floor, wavering with the rhythm of the tanks. "The town's uneasy. They've seen things. They hear things. You need to stop this before you bring ruin on all of us."

She closed her notebook slowly, pressing her fingers over the tremor in her injured hand before he could see. Her voice, when it came, was steadier than her pulse. "I'm not bringing ruin. I'm bringing restoration. The ocean is dying. If I can change these creatures, make them stronger, the seas have a chance."

Her grandfather stepped further in, setting the lantern on her workbench. Its warm glow clashed uneasily with the cold blue shimmer of the tanks. He studied her face for a long moment, as if searching for the child who used to chase gulls on the cliffs instead of the woman who stood here defying the tide.

His eyes softened, sorrow clouding them. "And if you've changed more than you can control?" His voice was low, edged with something heavier than anger. "You think you're saving the sea, but what if the sea no longer needs saving from us? What if it's preparing to save itself from you?"

Olivia's throat tightened, but she turned away before he could see the flicker of doubt. Her gaze settled on the tanks, where the jellyfish hovered in eerie synchrony. Their bells pulsed with quiet determination, their tendrils fanning outward in graceful waves, too graceful, too deliberate. They looked less like specimens now and more like an audience, listening.

The hum pressed faintly against her ribs, deep and patient, as if echoing her grandfather's words.

Olivia gripped the edge of the workbench until her knuckles turned white.

She had no reply.

Her grandfather let out a sigh, heavy as surf breaking on stone. He gathered his lantern, the flame sputtering again, and left her to the glow of her creation. The door closed behind him with a muted click.

Alone, Olivia stood before the glass, her reflection fractured by the crack that still streaked like lightning across its surface. The jellyfish pulsed once more, a rhythm that seemed to echo the beating of her own heart.

She did not answer him. But in the silence, she wondered if the sea already had.

Six

THE TRANSFORMATION

The welt on Olivia's wrist faded within days, but the memory of it did not. At night, when the waves crashed against the cliffs below, she often swore she could still feel the sting burning faintly beneath her skin, pulsing in time with her heartbeat. Sometimes it throbbed like a dull ache; other times, it flared sharp and sudden, as though the mark were not flesh at all but a hidden ember refusing to cool.

She told herself it was only nerves, just healing tissue and misfiring signals, nothing more. The human mind had a way of inventing ghosts, of clinging to what it feared. She repeated the words like a mantra when she lay awake, staring at the dark rafters above her bed and listening to the distant rhythm of the surf.

But the comfort never lasted long. The pulse would return, syncing with the tide, and she would find herself gripping her wrist in the dark, willing it still. Some nights, she swore the faintest glow seeped through the bandages, painting her skin in ghostly lines of blue. She would cover it quickly, burying her arm beneath the blanket, as if hiding it could make the truth less real.

And always, when the ache came strongest, she thought she heard it, the hum, softer than breath, curling in the silence of her room.

Yet when she returned to the tanks, her certainty wavered.

The altered jellyfish no longer drifted idly, no longer the lazy, uncoordinated wanderers she had cataloged in the weeks before. They hovered in tight formation, bells rising and falling as though they breathed with one shared lung. Each contraction came in flawless precision, each release perfectly timed, like the measured beat of a drum only they could hear.

Their tendrils swept outward in long, languid arcs, brushing the glass in unison. The motion rippled through them not as individuals, but as a collective, waves rolling across their translucent bodies, cresting, falling, repeating. It was hypnotic. Unnerving. Like a tide made flesh, bound not by currents but by something deeper, something chosen.

The longer she watched, the more Olivia felt her chest tighten. It wasn't randomness. It wasn't chance. It was choreography.

And though she tried to remind herself that they were only animals, she could not shake the growing dread that she was standing on the wrong side of the glass, that perhaps they were the ones observing her.

Olivia leaned closer, her breath fogging a pale bloom across the glass. The faint hum she had half dismissed before now surged with startling clarity, vibrating through the metal braces that held the tank in place. The sound wasn't just heard, it was felt, crawling through the bones of the room.

The shelves quivered. Beakers clicked together like teeth in a shiver, their thin rims ringing in uneven notes. The vibration ran through the stone floor and climbed up through the workbench until even the wood beneath her palm seemed to pulse with the resonance, warm and alive under her touch.

She pressed her good hand harder against it, as though anchoring herself, but the effort only made the tremor clearer. The hum wasn't background anymore. It had claimed the room.

Her wrist ached in sympathy, the scar glowing faintly as if answering the rhythm. The ache spread upward, coursing through muscle and bone, a phantom throb beneath her skin.

"Impossible," she whispered, but the word rang hollow, swallowed in the charged air.

She reached for her notebook. Her pencil shook so badly that the graphite carved erratic scars across the page, notes splintering into illegible fragments. Another pulse rippled through the tank, and the hum deepened. The beakers on the nearest shelf rattled in protest. One toppled, shattering on the stone floor with a crystalline scream. Shards skittered across the tiles like scattering ice.

Olivia flinched, and at that exact instant, the jellyfish surged forward. Tendrils struck the glass, lashing against the place her hand had hovered only a heartbeat earlier.

Her skin prickled. She froze. It wasn't random. It wasn't chance.

The creatures were not only aware of her, they were responding.

Her rational mind clawed for ground. Coincidence. Overwork. A tired mind seeing patterns where none existed. Yet beneath those excuses, something older stirred, an instinct deep as the tide: they knew her.

The hum swelled. It climbed higher and higher until it resonated in her lungs, her ribs, her skull, as though her very skeleton had been strung like an instrument. Every hair on her body lifted. Her teeth vibrated. Her heartbeat skipped, then fell into lockstep with the rhythm that filled the room.

She clutched the workbench, vision swimming as the glow intensified. Light spilled outward in waves, turning the air into shifting currents of green and blue. The walls dissolved into shadows of rippling water, the floor into a seabed of trembling light. For one disorienting moment, she was not in her cliffside lab at all. She was submerged, standing at the bottom of the sea, surrounded by something vast and alive.

And in that living light, she thought she saw them, shapes forming, dissolving, reforming. Not random flares but lines, arcs, symbols flickering too fast to fix yet too deliberate to dismiss. They weren't pulses. They were words.

Her notebook slipped from her numb hand, pages fanning wide as it struck the floor. Ink-smeared margins caught the glow, and for a terrible heartbeat, she swore her own scribbled notes shimmered back at her, as though her science had been read, recognized, and answered.

Her grandfather's voice whispered through the memory of countless evenings by the hearth: *And if you've changed more than you can control?*

The largest jelly pressed forward, slow and deliberate. Its bell contracted like a living heartbeat, tendrils unfurling until they spread across the glass. They traced her palm's outline as though memorizing it, mapping the lines of her flesh with impossible patience.

The hum softened, no longer deafening but intimate, its rhythm matching her heartbeat exactly. The lanterns dimmed and flickered, swaying into the same cadence until they guttered to pale ghosts of flame.

Olivia's chest rose and fell in shallow gasps. She lifted her trembling hand, the scar burning hot and bright, and laid it against the glass once more, not defiance, not control, but recognition.

The jelly pulsed in reply, slow, deliberate, certain.

The lab seemed to shrink around her, walls pressing inward as though the very stones were listening. The air grew dense, charged with a pressure that hummed against her skin. Every breath tasted of salt and iron. The windows fogged with condensation, rivulets running down their panes, though she knew the wind outside bit with winter cold.

Still, she could not look away.

Her scar burned like a coal, her heartbeat caught in the same rhythm as the light within the tank. The jelly pulsed again, patient and deliberate, as if offering proof of some unspoken kinship.

It's not just survival, she thought, her chest tight, fear and wonder twisting until she couldn't tell one from the other. It's evolution.

And for a terrifying instant, she felt as though she were standing on the wrong side of the glass.

Her hand trembled against the glass, caught between instinct and something far older than instinct. The jelly pulsed once more, steady as a drum, and the hum sank deep into her bones.

Then, mercifully, her knees wavered. The fragile connection broke. She stumbled back, dragging in a ragged breath as though surfacing from deep water.

The tank glowed on, serene and silent, its occupants drifting like nothing had happened. Only the faint scorch of her scar, glowing beneath her sleeve, bore witness.

Olivia pressed a shaking hand over it, forcing herself to breathe evenly.

The echo of that rhythm still lingered in her chest.

"Evolution," she whispered to the empty lab, the word tasting both sacred and damning.

And for the first time, she wondered if it was hers, or theirs.

ESCAPE

The storm hit at midnight. Wind clawed at the cliffs with a fury that made the old rock groan, rattling the shutters of Olivia's laboratory until the hinges shrieked like metal in pain. Rain slammed against the tall windows in sheets so thick that the world outside vanished into a blur of water and lightning. Each rivulet streaked down the panes, warping the glow from within the tanks until it seemed as if the jellyfish themselves were crawling free into the storm.

Below, the sea bellowed. Waves hurled themselves against the stone, their white crests torn to ribbons by the gale before they could fall. The thunder of water echoed through the cliffside, shaking the floor beneath Olivia's boots. Each strike of the tide seemed to answer the pulse of the creatures inside, synchronized, violent, and unrelenting.

Lantern flames guttered with every blast of wind that slipped through the seams of the shutters. The air was thick with salt, damp with spray forced in through cracks. Even the walls seemed to shiver, as though the sea's rage had seeped into the bones of the glasshouse itself.

And through it all, the tanks glowed steadily, uncaring, unyielding. Their light twisted with every rivulet of rain, painting the storm with ghostly streaks of blue and green, as if the creatures were bleeding their radiance into the night.

Olivia stood at her desk, scribbling notes by lantern light. The shadows of her hand jittered across the paper, warped by the uneven flicker of the flame. Her injured wrist throbbed with every beat of the hum, glowing faintly beneath her sleeve as if the creatures inside the tanks had reached into her very veins. She pressed harder with the pencil, the tip scratching deep into the page, forcing steadiness through sheer will. But the tremor betrayed her all the same.

The largest tank shuddered.

At first, she told herself it was the storm, the wind battering the walls, the waves hammering the cliffs, but then she saw the truth. Inside, the jellyfish had gathered in eerie formation, their bells flexing as one, tendrils coiling like drawn bowstrings. In perfect unison, they struck outward, slamming against the glass with a force that reverberated through the frame.

The pane groaned, a long, pained sound that rose above the howl of the gale. Tiny fissures spidered across the surface, branching outward like veins of lightning. Each fracture lit up with a ghostly blue glow, as if the cracks themselves had become conduits for the creatures' strange energy.

"No…" The word tore from her throat as her notebook slid from her grasp.

She lunged forward, fingers fumbling for the roll of sealing resin kept on the workbench for emergencies. Her hands shook as she slapped it against the spreading cracks, trying to seal the leaks, but the effort was like patching a dam with scraps of cloth. Water burst through in cold, relentless streams, soaking her sleeves, trickling down her arms, and puddling on the floor in quick, rhythmic ticks.

The hum spiked, vibrating against her ribs, filling her ears with a pressure that bordered on pain. The tank quivered under the strain. The resin patch bubbled and peeled away as if the glass itself were rejecting her efforts.

"Hold, please, hold, " she begged, pressing her weight against the pane as though her body alone might keep it intact. But beneath her palms, the fissures widened, glowing brighter, their branches stretching toward the edges of the tank like a spreading infection.

The hum rose to a deafening pitch, drilling into her skull until her teeth ached and her ribs felt ready to splinter from within. Every nerve in her body vibrated in tune with it. The lantern flame on her desk guttered wildly, casting grotesque shadows that leapt and twisted across the stone walls like writhing specters. The entire lab seemed to quiver on the edge of breaking.

Then the glass gave way.

The sound was not a clean crack but a detonation, an explosive roar, as though a cannon blast had been trapped too long within the pane. The shuddering release ripped through the air, rattling shelves, sending beakers tumbling, and papers fluttering like startled birds.

A wave of seawater surged outward, striking her with brutal force. The impact tore her from her footing and hurled her across the wet stones. She landed hard, the breath ripped from her lungs in a choking gasp.

Pain spiked through her hands as she scrambled up, the skin of her palms sliced by glittering shards that now littered the floor like jagged ice. Cold water poured around her ankles, rushing to find every gap, soaking the hem of her skirts, and clinging heavy as chains.

The tank loomed in ruin, its frame buckling, metal rivets groaning in protest. In the gaping breach, something moved, something more terrible in freedom than behind glass.

One of the jellyfish spilled free.

Its bell convulsed, furious and alive, as it rose from the wreckage like something unmoored from the deep. Tendrils dangled, trembling inches from her skin. Freed from its prison, it did not drift idly but moved with purpose, with a terrible elegance that belonged not to any sea creature but to something older, something elemental. Each pulse lit the storm-lashed lab in bursts of ghostly fire, walls, ceiling, and shattered glass all painted in rippling hues of green-blue radiance. Even her own trembling hands glowed as though they no longer belonged to her.

For the first time, Olivia saw it unconfined, and knew, with a sick clarity, how laughably fragile the glass had always been.

Her eyes darted to the emergency net by the door. She lunged, nearly slipping in the pooling water. Her hands shook so violently that the wooden handle almost slipped from her grip, the sodden mesh dragging behind like an anchor.

The creature pulsed once. Twice. Each flare of its bell was measured, deliberate, as though testing not the air but her resolve. Its tendrils reached downward, brushing the stone floor. Wherever they touched, faint trails of phosphorescence lingered, curling into strange, half-formed shapes that glimmered like fleeting script before dissolving back into darkness. For an instant, awe stole her breath, was it writing? Was it speaking?

Her hesitation cost her.

The jelly swept higher, its glow swelling until the air itself seemed to hum with it. Olivia forced her leaden limbs forward, teeth clenched.

With a ragged cry, she hurled the net upward.

The mesh caught across the edge of its bell. For a heartbeat, it seemed to hang suspended between them, between her frailty and its alien grace. Then the weight slammed into her. The net dragged the creature downward, but the force nearly toppled her with it.

Tendrils lashed like whips of fire. One grazed her shoulder.

Agony seared through her flesh, a brand of cold flame that spread instantly through her veins. The cry that ripped from her throat was half scream, half sob. The wound glowed fiercely beneath the torn fabric, bright enough to cast her shadow against the wall.

Still, she clung to the net, knuckles white, every nerve screaming. She would not let go.

Pain seared through her body, blinding and absolute, her vision exploding into white. A scream tore from her throat as she staggered, muscles convulsing. The air itself seemed to vibrate with her cry, as though the creature had made her body another instrument for its hum. But she held on.

Her arms trembled, her shoulders burned, yet her grip did not loosen. With every ounce of strength she had left, she forced the net downward, dragging the thrashing, luminous weight toward the fractured tank. Seawater sloshed around her calves, soaking her boots; glass cut deeper into her palms with each desperate tug. The hum rattled through her bones until her teeth ached, until her pulse itself seemed to beat in time with the alien rhythm, as though she were no longer fighting the thing, but bound to it.

At last, with a convulsive shudder, the jelly slid back into the water. The sound it made was not a splash but something softer, almost sentient, a sigh, a note of release that rippled through the ruined lab. The hum broke off abruptly. Silence crashed down, sudden and suffocating, broken only by the ragged hitch of Olivia's breath.

Her knees buckled. She dropped to the wet stone floor, the net slipping from her hands. For long moments, she could only kneel there, chest heaving, her body trembling with aftershocks of pain. Her shoulder still burned where the tendril had grazed her, the welt glowing faintly, each throb answering the faint pulse of the jellies still hovering within the tank.

The lantern guttered low, its flame pitiful beside the cold fire filling the room. The lab smelled of brine, blood, and scorched iron.

Olivia wrapped an arm around herself and forced her eyes toward the tank. The jelly hovered within, calmer now, its bell pulsing in measured intervals, tendrils drifting with deceptive gentleness, as if satisfied, or watching.

She whispered hoarsely, to no one but herself:

"It isn't caged. None of this ever was."

The moment its bell submerged, the hum broke off abruptly. Silence crashed down, heavier than the storm outside. The only sounds were the thunder overhead and Olivia's own ragged gasps.

She collapsed to her knees, chest heaving, sweat mingling with seawater on her skin. The wound on her shoulder glowed faintly, a sickly blue that pulsed in rhythm with the remaining creatures in the tank. She pressed her palm against it, but the light bled between her fingers, stubborn as fire beneath ash.

For long minutes, she sat there, drenched and trembling, her hair plastered to her face, staring at the rippling glow inside the ruined tank. The storm rattled the shutters, water dripped steadily from the broken glass, and yet her eyes did not stray from the jellies. They pulsed slowly, deliberately, their light brushing over her skin as though the tank itself had become a window rather than a barrier.

Her mind screamed to abandon the experiment. To call for help. To flee the cliffside laboratory and never return. She pictured herself at the harbor, running barefoot across the slick boards, collapsing into her grandfather's arms, confessing everything. She pictured the villagers' faces when they learned. She pictured Alex's quiet stare, the one that always seemed to measure more than he said aloud.

But her hands, almost of their own accord, reached for her notebook lying sodden on the floor. Its pages clung together, edges curling, ink bleeding like veins through paper. Still, she forced it open, fingers numb and shaking, and scrawled across the damp page:

Specimen breached containment. Subject responsive to environment outside tank. Highly coordinated. Possibly… sentient.

The words bled as the paper drank the water. Her pencil snapped in two, the sound sharp in the hush.

Olivia stared at the broken halves in her fist, her chest heaving, the glow in her shoulder still burning through the fabric of her sleeve. In the silence that followed, a thought pressed against her mind, as cold and inexorable as the tide:

It knows me now.

Olivia lifted her head toward the rain-lashed window. Far below, where the waves battered the cliffs, she thought she saw faint lights flickering beneath the black water. Not lanterns. Not lightning. A pulse, steady and deliberate, echoing from the deep.

Her breath caught in her throat. The storm rattled the glass, thunder rolling low across the horizon, but she hardly heard it. All she could see were those lights, flaring in sequence, vanishing, then blooming again, like a heartbeat rising from the abyss.

She pressed her good hand to the pane, rain running cold down the outside while sweat burned on her skin. The reflection that met her eyes was pale, wild, and rimmed with a ghostly glow from her wounded shoulder. For a heartbeat, she did not know if it was her own image or something watching from the other side.

She whispered into the storm, the words barely more than breath, carried away by the wind:

"They're already in the sea."

As if answering her confession, the hum returned, so faint it might have been mistaken for the gale, but no less real. It wound its way through the beams and floorboards, a subtle vibration that bypassed her ears entirely and settled straight into her bones. Her stomach turned cold.

This time, it did not come from the tanks.

It came from the ocean itself.

Eight

AFTERMATH IN THE LAB

The storm had passed, but the lab still looked as if it had been submerged. Salt water pooled across the floor in uneven patches, dark where it gathered deepest, and glinting faintly where the lantern flame caught on ripples. Shards of glass floated like broken stars in the puddles, shifting with each faint draft that crept through the cracks in the window frames.

Papers lay plastered against the stones; their neat lines of ink blurred into gray stains that spread like bruises. Charts she had labored over for weeks clung to the floor as if the sea had claimed them, their edges curling and tearing under her trembling hands as she tried to lift them.

The air was heavy with the stench of brine and copper, sharp as blood, metallic as rust. It coated her tongue, filled her lungs, and made her sanctuary feel less like a place of discovery and more like a grave. The sea had forced its way in, and with it had left a reminder: nothing she built here was truly hers to control.

Olivia crouched in the wreckage, numb hands gathering the scraps of her work. Pages peeled off the stones with a sickening sound, their ink dissolving into gray smudges that stained her palms. She wrung them out one by one, but each effort only smeared her words further, as if the storm itself mocked her attempts to salvage what was already lost. Weeks of observations, hours of sleepless nights bent over glass and flame, all destroyed in a single crack of glass.

The smell of salt and wet paper filled her nose, clinging to her hair and her skin. She stacked the ruined notes in a trembling heap, the pile collapsing under its own weight like ash.

Her arm ached, a low, steady burn that had become too familiar in the past days, as if the storm had only deepened it. With reluctant fingers, she unrolled her sleeve.

She winced. The welt had not faded. Instead, the mark had darkened into something unnatural, a raised ridge that glowed faintly from beneath the skin. Sickly blue radiance pulsed in rhythm with her heartbeat, a dim lantern beneath her flesh. Each throb was a whisper, a reminder, sending shivers across her chest as if the hum of the creatures still lived inside her bones.

She pressed her sleeve down quickly, but the light seeped through the linen, bleeding into the dim air. For a moment, she imagined the entire room pulsed with it, the drowned floor, the broken glass, even the air itself responding to that rhythm.

Olivia curled her fingers tight around her arm, as though pressure alone might silence it. But the pulse only answered more strongly.

She pressed her sleeve back down, but the glow bled through the fabric, staining it with pale fire. No matter how tightly she bound it, the light refused to dim. A sound behind her made her flinch, the creak of the lab door.

"Olivia?"

She turned, startled. Alex Shepherd filled the doorway, lantern in hand, his fisherman's coat dripping from the storm. Water ran in rivulets from the hem, pooling at his boots. His face, half-lit by the lantern's glow, was carved with fatigue and something sharper, concern. His gaze swept over the wreckage: the drowned floor, the shattered glass, the tank still trembling faintly as though alive.

"You shouldn't be here," Olivia snapped, too quickly, her voice frayed by exhaustion and fear.

Alex didn't move. His shoulders filled the frame of the doorway, broad and immovable, as though the storm outside had carried him here with purpose. He stepped forward, the lantern's light catching in his eyes.

"You're hurt."

The words weren't a question. His gaze had already fixed on her wrist; on the way she clutched it close against her chest as if guarding a secret.

"It's nothing," she said. The lie cracked under the weight of her own unsteady breath.

In two strides, he was before her, boots sloshing through seawater. He reached for her arm. Olivia stiffened, but his touch was gentle, careful in a way that disarmed her more than any sternness could. His hands were calloused from years on rope and net, yet they moved with surprising tenderness as he coaxed her sleeve back.

The glow answered him. It flared at his nearness, as though her pulse had quickened of its own accord, or as though the thing in her veins recognized him. For an instant, the light painted both their hands, his rough skin rimmed in cold fire.

Olivia tried to pull away, but her strength failed. "It's not, it isn't like any injury you've seen before."

"Doesn't matter." His voice was low, steady, more like the sea at slack tide than the storm raging beyond the cliffs. He guided her to the workbench, where the lantern's glow pooled across the battered wood. With the sweep of his arm, he cleared a space, scattering sodden papers to the floor, and set the light down.

From his coat pocket, he drew a small tin of salve, the kind carried by every fisherman to ward off rope burn and salt blisters, and a scrap of clean cloth wrapped carefully against the damp. "Hold still."

Olivia opened her mouth to argue, to tell him he couldn't understand, that this was beyond fish cuts and dockside wounds. But the words faltered on her tongue. Her arm trembled in his grip, and with it, her resolve. She let him press the cloth to her skin.

The sting flared white-hot. Her breath hitched, teeth clenching against a cry. Yet the steadiness of his hands anchored her, as if the heat of the glow could not fully reach her so long as his touch remained. He worked in silence, jaw set, rain and wind filling the spaces between them.

Her eyes flicked to his face. Drops of seawater clung to his hair, catching the lantern light like beads of glass. He did not look at her, only at the wound, as though nothing else in the world mattered but setting it right.

For the first time since the tank had shattered, Olivia felt the pounding of her heart slow. The hum was still there, deep in her bones, but Alex's presence dulled its edge.

Her half-soaked notebook waited, pages curling where ink had bled into the grain. The scrawled words glimmered faintly in the lantern glow, as though the sea itself had stained them.

When Alex bound her wrist with practiced efficiency, Olivia found her gaze wandering from his hands to his face. He worked with the same patience she'd seen him use untangling snarled nets on the docks, calm, methodical, as if no task was too ruined to save. His fingers, calloused and strong, were surprisingly gentle as they wrapped the cloth snug around her arm.

She caught herself studying the lines of concentration carved into his brow, the steady set of his jaw. The lamplight traced the faint scar that ran from his cheekbone to the edge of his jawline, a mark she'd noticed before but never asked about. In the storm's flickering shadows, it looked less like an old wound and more like a badge, a quiet testament to battles endured.

For a moment, the ruined lab, the shards of glass glittering like ice, the endless hum vibrating in her bones, all of it receded. There was only Alex, close enough that the damp scent of salt and rain clung to him, steady enough that her trembling stilled.

"Thank you," she whispered, the words fragile, almost drowned by the storm raging against the windows.

Alex's eyes flicked to hers only briefly, but the weight of that glance pinned her breath in her chest. He didn't linger. His focus returned to the knot he tied, pulling it firm but not cruel.

"You can't keep doing this alone," he said, voice low, almost rough. The words weren't a scolding, they were too quiet, too weighted with something else: concern, resolve.

Olivia's throat tightened. She wanted to protest, to insist she could handle it, that this was her burden, her creation, her mistake to mend. But the way he said it left no room for argument. He wasn't asking. He was stating a truth.

The hum swelled faintly, pressing against the walls. The tanks flickered with ghostly light, but for that small stretch of time, Olivia's fear was less sharp. Alex's presence steadied it, as if the storm outside could rage, the sea could rise, the jellyfish could pulse their strange signals, and still, here at her side, he anchored her.

When he left a few minutes later, lantern glow trailing into the storm-muted night, Olivia sat frozen, staring at the bandage. His careful knotting held firm, neat and certain in a way her own trembling hands could not have managed. Yet the glow still bled faintly through, pulsing with its unnatural rhythm, muted now, softened by his care, but no less present.

The lab groaned around her, beams settling after the storm, water dripping steadily into the pools on the floor. Each drop sounded louder in the silence Alex left behind, as if the world had shrunk back to its rhythm of damage and waiting.

At her desk, she pulled her notebook from the pile of damp papers. Half the pages were ruined, ink leached into gray fogs, words lost forever. But a handful remained dry enough to take her hand. She turned to one and forced the pencil across the page, each letter a battle against her fatigue:

Subject breached containment.

Injury sustained.

Persistent luminescence.

Correlation between wound and resonance of specimens.

Possible permanent alteration.

The graphite dragged unevenly; her hand shook, her wrist ached. When she stopped, the words swam before her eyes, dissolving into meaningless shapes. A brittle sound escaped her throat, too sharp to be a sob, too hollow to be laughter. She clamped a hand over her mouth, but it echoed anyway in the ruined lab.

Permanent alteration.

She had wanted resilience for the ocean. She had wanted to save reefs, to make the seas stronger, to give something back to the only world she'd ever truly loved. Instead, she had rewritten herself, her body carrying the proof of her meddling like a brand.

Her gaze shifted to the darkened window. Outside, the storm had passed, leaving only the sea's restless churn below. But the glass reflected her own face more than the view. And for a heartbeat, she did not recognize it.

Her skin was too pale, her eyes ringed with shadows, no, with light, faint halos that shimmered in tune with her heartbeat. Her shoulder, beneath the damp fabric, glowed steadily, the bandage a thin veil that could not contain it. The reflection shimmered, as though something beneath the surface pressed forward, waiting to emerge.

She flinched back, turning away sharply, breath ragged. Her pulse roared in her ears. For a moment, she clutched the edge of the workbench just to steady herself. The hum lingered low in her bones, reminding her that she was no longer only observing her specimens.

She was becoming one of them.

She crouched before the hearth, the ruined pages heavy in her arms. Ink smeared her palms black, mingling with blood from shallow cuts until her hands looked like they belonged to someone else entirely. The paper sagged with damp, fragile as skin, threatening to tear apart before she even reached the fire.

One match. That was all it would take. One hiss of sulfur, one spark, and the work that had consumed her sleepless nights, the work that had cost her blood and perhaps her humanity, would be ash.

Her fingers found the matchbox on the mantle. She turned it over in her hand, the rasp of wood against cardboard loud in the silence. The creatures hummed louder behind her, the tanks vibrating softly as though they too knew what she was about to decide.

Olivia's hand shook. She imagined the flames catching, imagined the relief of watching the words curl into smoke and vanish. No one could blame her then. No one could point to her notes and say this was the blueprint of their ruin. No one would ever know what she had done. But the thought curdled almost as quickly as it came. If she burned it all, she burned the only map through this nightmare.

Every line she'd written was knowledge, dangerous, damning, incomplete, but knowledge nonetheless.

And without it, Haven's Reach would face the tide blind.

She closed her eyes; the matchbox pressed to her forehead. The faint glow of her wrist lit the damp pages in her lap, a ghostly lantern, reminding her that she carried the evidence whether she wanted to or not.

Her grandfather's voice echoed in memory, worn with age but steady:

The sea doesn't yield to fire or fury. It yields to understanding. Her hand lowered. She placed the matchbox back on the mantle with deliberate care. Then, with her uninjured hand, she drew a single dry page from the wreckage, set it on the workbench, and picked up her pencil.

Her wrist burned, the hum pressed close, and still she wrote, her letters jagged but defiant:

Containment impossible.

They are more than an experiment.

They are aware.

The pencil snapped in her grip, but she didn't stop staring at the words. For the first time since the glass shattered, she knew there was no turning back.

Her whisper vanished into the air, devoured by the thrum of the tanks. The jellyfish drifted close, their tendrils brushing the glass as though straining to hear her vow. Their glow flickered brighter for a breath, then dimmed again, like an acknowledgment, or a warning.

Olivia pressed her aching wrist against her chest, feeling the faint tattoo of light pulsing beneath her bandage. It beat in time with the creatures, a tether she could not sever. The matchbox still lay where it had fallen, its dull tin catching the glow like a half-buried relic in sand. Her gaze lingered on it. Burning everything would have been simple. Clean. But nothing about this was simple anymore. She had chosen the harder path, the only path she could live with. She slid the folder shut, her hand still trembling. The paper crackled like brittle ice, fragile, but it was all she had: a record of what the town would never believe and what she could not deny.

Lightning split the horizon, briefly painting her reflection in the windowpane: pale face, hollow eyes, and that faint, unearthly glow seeping through her sleeve. Not just a scientist. Not just a woman.

Something caught between.

Her jaw tightened. "I will fix this," she said again, louder this time, as though daring the sea itself to hear her.

The hum receded into a low, steady drone, almost like a heartbeat.

And for the first time that night, Olivia realized she no longer knew whether it was hers or theirs.

Nine

THE GRANDFATHER'S
VISIT

The knock came just before dawn. Olivia froze where she sat hunched at her desk, her body aching from hours locked in the same position. Ink smeared her fingertips, black crescents under her nails, and the page before her blurred into gray smudges. Beneath her bandage, her wrist still glowed faintly, a small coal burning stubbornly in the dark. The light seeped through the linen in thin lines, pulsing as though the creatures in the tanks had found a way to mark her as one of their own.

She hadn't slept. Every attempt had been broken by the memory of the shattered glass, by the phantom echo of water rushing over her skin, by the steady hum that seemed to linger even when the room was silent. The tanks behind her swelled and dimmed with their rhythm, the glow rising like breath in a chest, each pulse heavy as a heartbeat.

At first, she thought the sound was her imagination, the dull rhythm of exhaustion making ghosts out of every creak and sigh. But it came again: knuckles on wood, muffled by the storm-wet wind.

Her chair scraped against the stones as she rose. She pulled her sleeve down hastily, hiding the faint shimmer beneath the cloth, then shoved her notebook beneath a stack of sodden papers. The motion was frantic, her hands clumsy from fatigue, as though burying the evidence could undo the night before.

The knock sounded again, louder this time, insistent.

Olivia crossed the room in quick strides, her bare feet splashing through shallow pools where the floor still held seawater. She unlatched the swollen door, the wood groaning in protest, and pulled it open.

Her grandfather stood in the frame, lantern in hand. Rain had plastered his silver hair to his head, and the hood of his oilskin cloak dripped steadily onto the threshold. His lined face was carved deep by worry, but when his eyes fell on her, on the slump of her shoulders, the dark hollows under her eyes, the tremor she couldn't quite hide, his sternness softened, just for a moment.

"Child," he said, his voice low and steady, carrying the weight of both rebuke and concern. "You look worse than the storm that passed."

Olivia tried to summon a smile, some shield against the pity in his eyes, but it faltered before it could form. "I'm working," she murmured.

Her throat felt raw. "You shouldn't be out here."

He stepped inside without waiting for her protest, brushing past her gently with the care of a man who had raised her, scolded her, protected her. The lantern swung at his side, casting arcs of amber light across the wreckage. His gaze swept the room slowly, taking in the shattered glass that had been swept into uneasy piles, the salt stains white against the dark stone floor, the tanks still glowing with restless life as though mocking her attempts at order.

The lantern light caught the fragments scattered near the hearth, tiny shards that gleamed like stars spilled across the floor. He paused, his breath catching. "God above," he whispered, not in anger, but in awe edged with dread. His voice trembled with the weight of seeing proof of what he had feared but not spoken aloud. "What have you done?"

Olivia stiffened, her back straightening as though she could brace herself against his judgment. The shame that gnawed at her twisted tighter, but another instinct, defiance, rose with it, sharp and unyielding.

"I was trying to save them," she said, her voice louder than she meant, carrying too much desperation to sound steady. "To save all of it. The reefs, the fish, the ocean itself."

He turned to her then, the lines of his face deepening with sorrow. In the lantern's flicker, his eyes looked darker, shadowed by both love and fear.

"And instead, you've lost something on this town," he said. His words landed heavy, like stones cast into deep water.

"The people talk, Olivia. Lights on the cliffs, humming in the night, animals gone missing. They're frightened. And now," He glanced at the glowing tanks; their pulsing light reflected in his weary eyes. "Now I see why."

Her throat tightened until it hurt to breathe. She reached for the edge of the workbench, fingers splayed against the damp wood, steadying herself against the surge of shame.

"I didn't mean for this to happen," she said at last, her voice breaking under the weight of truth. "I thought…"

Her words faltered, swallowed by the hum that seemed to stir faintly at the edges of the silence, as though the creatures themselves waited for her to finish.

"… I thought I could control it." The words left her lips in a whisper, brittle and sharp, and conjured an image unbidden: Alex Shepherd's hands binding her wrist, the pressure of cloth firm but gentle, his voice steady as he told her she couldn't keep doing this alone. Heat rose to her cheeks before she could stop it. She pushed the memory down as quickly as it surfaced. This was not the time. Not here, not with her grandfather's eyes fixed on her.

He stepped closer, setting his lantern on the table. Its golden light spilled across the wreckage of glass and ink-stained pages, mingling uneasily with the cold glow that pulsed from the tanks.

"The sea doesn't answer to control," he said softly, but his words carried the weight of years, of storms weathered and tides endured. "You were taught that. Respect, balance, not this."

He gestured to the tanks, where the jellyfish pressed against the glass in eerie unison. The light spilling from them throbbed like a warning.

Olivia bit down on her lip until she tasted copper.

"They're in the ocean already," she confessed, the words scraping her throat raw. "One escaped during the storm." Her voice cracked, her shoulders sagging as if speaking it aloud made it real again. "I can't take it back."

Her grandfather's silence was long and heavy. Only the hum filled the space between them, steady, patient, a reminder that time was no longer hers to command. At last, he sighed, the sound weary as wind across the cliffs. His shoulders drooped under the burden of years and worry.

"Then we'll have to find a way forward."

Her head snapped up, eyes wide. "You mean,"

"I mean you're not alone in this." His gaze softened, though his voice held its steel. The same compassion that had once guided her small hands to trace the names of sea stars in tide pools shone in his eyes now.

"You've gone too far, Olivia. But you're still my granddaughter. If there's a way to mend this, we'll mend it together."

Her throat closed around a sob she refused to let out. Tears blurred her vision, making the lamplight smear into molten streaks. She swallowed hard, the lump in her throat unyielding.

Her grandfather reached into his coat and drew out a weathered leather journal. Its edges were warped by salt and age, the leather scarred with decades of use. He pressed it into her hands with a gravity that spoke of inheritance, not just offering.

"Your science can tell you how," he said.

"But sometimes the old ways remember why. Between the two, perhaps we can find a cure."

Olivia clutched the journal to her chest as if it were a lifeline. Its weight was solid, grounding, the way Alex's hands had steadied her when her own trembled too violently to hold a bandage. The memory of that moment returned despite her best efforts, lingering this time not as heat in her face but as a faint warmth that reached past the fear. Between her grandfather's strength and Alex's quiet, unshakable presence, there were still anchors tying her here, keeping her from being swept away entirely by the hum.

The tanks pulsed brighter, as though mocking her fragile hope, but she forced herself to meet their glow with her chin lifted. For the first time since the glass had shattered, she did not feel entirely alone.

Ten

RESEARCH & BONDING

The days that followed blurred into a rhythm of work. Olivia cleared the wreckage of her ruined lab while her grandfather brought order to the chaos with the patience of age. Together they laid new boards across the cracked floor, sealed the leaking tanks, and mended what they could. The hum of the jellyfish remained constant, an uninvited reminder of their unfinished task, vibrating faintly through the walls no matter what repairs they made.

By day, Olivia bent over her microscopes, peering at tissue samples she had scraped from the damaged tank. She tracked cell growth, venom potency, and the strange luminescence that refused to die even outside the host. Her notes grew dense, margins filled with equations, sketches, frantic corrections, some written with such speed the graphite tore through the paper.

By night, her grandfather sat opposite her at the long worktable, his weathered hands turning the pages of old journals and sailors' logs. He spoke softly of herbs gathered from tidal pools, of barnacle salves once used to ease stings, of plants said to carry the sea's own healing. He lit the lantern low and spread leaves and roots across the table as though arranging artifacts, each placed with reverence.

Now and then, Alex Shepherd appeared at the doorway, as though the storm had left behind the habit of checking in. Sometimes he carried small things from the market, a bundle of fresh bread wrapped in cloth, dried fish, a flask of tea from the tavern, set down quietly on the counter before anyone thought to ask. Other times he stayed only long enough to exchange a word with her grandfather, his low voice rumbling like distant surf, before making his way back down the narrow cliff path.

Olivia never asked him to stay, yet each visit tugged at the air like a tide pulling against her chest. She would hear his boots fade down the stones and find herself staring at the door longer than she meant to, her pencil poised useless above the page.

And always, the hum lingered. Whether they worked in silence or spoke in fragments, of science, of old remedies, of the sea's mercurial moods, it threaded through their days like an uninvited witness, waiting.

At first, their approaches clashed.

"Superstition," Olivia muttered one evening, brushing aside a sprig of dried kelp her grandfather had laid near her notes. The fronds left a faint smear of salt across the paper, staining the edge of her careful sketches. "We need measurable compounds, not fishermen's tales. Kelp and stories won't stop what's coming."

"And yet," he replied gently, sliding the kelp back toward her with a steady hand, "the old ways kept them alive when the sea turned against them. Before pumps and microscopes, they listened. To tides, to plants, to patterns written in the moon. You'd do well to respect what endured."

Olivia opened her mouth to argue, then closed it again. The retort burned on her tongue, but shame tugged at her heart. Memories came unbidden, being a girl at his side, wide-eyed as he pointed out starfish in tide pools, explaining how each arm grew back when broken, how each creature had its place. She had believed him then, trusted balance more than boundaries. Before the years inland, before the books and professors had taught her to see the sea as fragile, failing, in need of mending by clever human hands.

So she let the kelp stay.

Later that night, under the wavering lamplight, she dissolved the brittle fronds in solution, more to humor him than from conviction. Yet when she placed a drop beneath her slide and peered into the eyepiece, her breath caught. The jellyfish cells, usually restless, their glow an untamed flicker—recoiled at the contact. The light dimmed, faint tendrils shrinking as though in retreat.

She blinked, adjusted the lens, repeated the test. The result was the same. Her pencil hovered useless above the page, then scribbled frantically, numbers and sketches jostling for space in the cramped margin.

When she looked up, her eyes wide, her grandfather was already watching her, his face half in shadow, half bathed in the blue glow of the tanks.

"You see?" he said softly, his smile weary but warm. "Science and superstition are sometimes only different languages for the same truth."

The days grew longer, the nights colder, but the rhythm carried them forward like a tide. Olivia adjusted oxygen levels in the tanks, her fingers steady on the valves, while her grandfather charted the moon's phases with patient diligence. He insisted the sea answered to more than salinity and temperature, that its moods bent to the pull of silver light. She laughed at his stubbornness at first, teasing him about sailors' tales, but when her samples pulsed brighter during the full moon, when the glow itself seemed to sharpen as though under lunar command, she stopped laughing.

Sometimes, Alex Shepherd came by with supplies. He was always practical about it, never lingering longer than he felt necessary. A bundle of planks for the floor, nails wrapped in oilskin for the broken frames, a coil of rope to reinforce the windows, or fresh fish set quietly on the counter when he thought Olivia had forgotten to eat. He said little, never wasting words. A stiff nod, a muttered, "Storm's easing," or "The nets are running light," was all he offered before turning back to his work.

Her grandfather welcomed him readily, as he always had, clapping him on the shoulder, thanking him, drawing him into brief conversation about tides or timber. Alex listened with the patience of someone who had been raised more by silence than by voices. Yet Olivia noticed the subtle tells: the way his jaw tightened when he passed the glowing tanks, how his shoulders squared whenever she flexed her bandaged wrist. It was not disinterest, she realized, but vigilance. His silence was not emptiness, but restraint.

Once, when a waterlogged board slipped from her grasp, Olivia nearly dropped it on her foot. Alex caught it with one hand, his grip unshaken despite the wood's sodden weight. He set it in place without comment, his expression unreadable. She thanked him, her voice coming out more awkward than she intended. He only gave a low grunt in reply—but the flush that crept across his ears in the lantern light betrayed him more than words ever could.

Olivia turned quickly back to her notes, pretending not to notice, though she found herself smiling faintly, despite the heaviness of the room. The jellyfish pulsed behind the glass in perfect synchrony, their glow reflecting on Alex's broad shoulders as he hammered the board into place. In the strange stillness of the lab, it struck her that not all strength was loud, and not all silence was distance.

By the end of the week, their work bore fruit: a serum brewed from crushed ocean plants, distilled chemicals, and fragments of jellyfish tissue tempered in saltwater. It shimmered faintly in its vial, a liquid that seemed to glow from within, as though holding a sliver of the sea's heartbeat.

Olivia held it up, her hand trembling despite herself. For the first time in days, the tightness in her chest eased, a breath of fragile hope slipping free.

"This could work," she whispered, afraid that saying it aloud might break the spell.

Her grandfather set a steadying hand on her shoulder, his calloused palm grounding her. "It must," he said simply.

His voice carried the weight of years, of storms weathered and battles endured, and in it she found the faintest comfort.

In the corner, Alex adjusted the lantern wick. The flame flared and steadied, shadows stretching and folding across his broad frame. He said nothing, He rarely did, but Olivia felt the weight of his presence as keenly as the hum from the tanks. When her hand shook, she caught the barest flicker of his gaze on her. Not critical, not curious, but steady. Watching, the way he always did when he thought she wouldn't notice.

Her pulse quickened in response. She lowered the vial slightly, pretending to study the light within, though her thoughts had tangled elsewhere. Alex had always been a man of silence, a man the town called aloof. Yet in that silence there was something unspoken, a watchfulness that held her steady in ways she could not name.

The jellyfish pulsed in their tanks, the glow answering faintly to the vial in her hand. Behind her, her grandfather spoke again, voice low and certain:

"What's made can always be unmade, child. Remember that."

But Olivia's mind lingered on Alex's gaze, on the way it steadied her more than the words did. For the first time since the glass had shattered, she felt the faintest shift in the balance, as if she wasn't carrying the weight alone.

Eleven

TRIAL OF THE SERUM

The sea was calm when they set out, but Olivia did not trust it. The calm was a mask, stretched too smooth across the water, hiding what waited beneath. She could feel it in her wrist, the scar that pulsed faintly beneath its wrappings, a quiet throb that kept time with the oars. It was less a wound now and more a warning, a reminder etched into her skin.

Every beat of her heart seemed to echo with the memory of the hum. Sometimes it was faint, like a half-forgotten melody, but now it lingered close, just at the edge of hearing. The stillness of the horizon only sharpened her dread. The sea did not rest. It never rested.

She sat rigid in the bow of the boat, lantern light painting her profile in fractured gold, watching the black water ripple under the blades of the oars. Each stroke carried them further from Haven's Reach, further into the expanse where the hum had first begun.

Her grandfather rowed with the strength of habit, every motion steady, though his eyes scanned the horizon as if he too knew peace was a lie. Between them, the crate of vials hummed faintly, its glow a ghostly heartbeat against the dark planks.

Alex sat in the stern, silent as stone, one hand resting on the coil of rope at his feet. He hadn't argued when her grandfather asked him to come, hadn't spoken at all, really, except for a curt, "I'll be there." Now, his broad frame was a shadow against the lantern light, shoulders squared as though he could shield them by his presence alone.

He rarely glanced at her, but Olivia felt the weight of his awareness all the same. When her hand trembled against the crate, his gaze flicked to it, quick, steady, grounding, before shifting back to the horizon. He said nothing, but the message was clear: I see you. I will not let this boat go down easily.

Olivia pressed her palm more firmly to the lid of the crate, not for balance, but as if she could will the vials' light to stay steady. Her wrist flared beneath its wrappings, heat pulsing through her veins. She closed her eyes for a moment, breathing in the brine, and whispered to herself:

"The sea remembers."

She sat at the bow of the small wooden boat, clutching the crate that held their vials of serum, while her grandfather steered them toward the open water. The lantern they carried barely pierced the night, its glow swallowed by the horizon. The waves slapped against the hull in a rhythm that was almost too even, as if the sea itself were breathing beneath them. The crate vibrated faintly against her knees, the glow inside bleeding through the cracks in pale threads of light. Each lurch of the boat sent a faint shimmer across her bandaged wrist, as though the serum itself responded to the scar. Olivia tightened her grip until her knuckles blanched.

Alex was there too, rowing with silent strength. He had insisted, not with words, but by appearing at the dock when they pushed off, a coil of rope slung over one shoulder, eyes dark with quiet resolve. He hadn't explained why, and Olivia hadn't asked. It wasn't his way. He stood apart from the others in Haven's Reach, always had. Yet here he was, shadow and muscle made flesh, rowing into the dark as though the act itself was inevitable.

Part of her wanted to tell him to leave, to stay away from what she had unleashed. To go back to the safety of the nets and the harbor where he belonged. But when he took his place at the oars, his broad hands curling around the worn wood, his movements steady as the tide itself, something inside her loosened.

The storm had left her brittle, her strength stretched thin by nights of sleepless vigilance, but Alex's presence settled like ballast in her chest.

For once, she wasn't entirely afraid of capsizing.

The silence between them was thick, broken only by the rhythm of the oars and the hiss of water against the hull. Her grandfather muttered low now and then about the tides, about how far they would need to go. But Alex said nothing. Still, when she risked the briefest glance back, she caught the faintest glimmer of his eyes in the lantern glow, watchful, steady, as though measuring every shiver of her breath.

The hum came first, faint but unmistakable. A low vibration that trembled in the air, in their ribs, in the boards beneath their feet.

Then came the light.

Dozens, no, hundreds, of jellyfish pulsed just beneath the surface, their bells glowing like drowned stars. Tendrils writhed and unfurled, some brushing the underside of the boat. The water shimmered with their radiance, casting ghostly patterns across Olivia's pale face.

Her grandfather spat over the side. "The sea doesn't welcome us tonight."

Olivia's pulse hammered. She uncorked the first vial. The liquid shimmered faintly, glowing in sympathy with the creatures below. Her hands shook as she rose, balancing herself against the rocking of the boat.

Alex's eyes flicked to her, quick and sharp. "Steady," he muttered, his voice low. Not tender, but firm, a command meant to anchor her.

The jellyfish surged upward, bells breaking the surface, tendrils lashing at the air as though reaching for them. One struck the hull with a wet slap, leaving a streak of blue fire. The boat shuddered.

"Now, child!" her grandfather barked.

Olivia hurled the vial into the water. Glass shattered, and the serum bloomed outward in a radiant cloud. For a heartbeat, nothing changed.

Then the hum faltered.

The jellyfish closest to the boat recoiled, their glow dimming as they drifted back. The cloud spread wider, and where it touched, tendrils shrank, curling in on themselves like burned paper. "It's working," Olivia whispered, clutching the gunwale. "It's,"

A tendril whipped out of the water, striking across her sleeve. Pain flared up her arm, white-hot. Her knees buckled, and the crate jolted dangerously at her side.

"Olivia!"

Alex was there in an instant, his arm locking around her waist before she toppled. He caught her with a strength that pulled her against him, steadying both her and the box at once. The boat rocked hard, but his grip held firm, braced like iron. Her breath hitched against his chest, the hum vibrating between them. Her grandfather leaned forward, gnarled hands steadying the crate where Alex held it tight. His weathered face was grim, but his eyes burned with urgency. "Hold her steady," he growled at Alex, before snatching up another vial with his free hand. "This is no time to hesitate." He flung the vial wide. The sea hissed where it landed. More jellyfish convulsed, their light sputtering. The hum fractured into a broken chorus, jagged and discordant.

But the swarm pressed closer still, a wall of pulsing bells and writhing tendrils. Their glow lit the night like cold lightning, bleaching the sea to silver and painting the three figures in unearthly hues. Each pulse felt heavier than the last, as though the ocean itself had grown a heartbeat. Olivia forced herself upright, though her legs shook beneath her. The crate loomed at her side, each vial shimmering like captured stars.

Her arm throbbed where the tendril had struck, the welt burning blue beneath the torn fabric, spreading its glow like veins of fire.

Her grandfather's voice cut sharp across the water. "Now, girl!"

Olivia clenched her teeth and raised a third vial. Her fingers trembled, slick with seawater, but she refused to let go. With a desperate cry, she smashed it against the side of the boat. The glass broke in a clean, ringing crack. Serum spilled in radiant streaks that slid into the sea like ribbons of molten light. The effect was immediate. The nearest jellyfish convulsed, their bells contracting violently.

The swarm shrieked, not with sound, but with vibration so raw and violent it rattled the oars, bent the wooden ribs of the boat, and set Olivia's teeth on edge until her jaw ached.

Then, one by one, the creatures began to sink. Their glow dimmed as though a current pulled it from them, their tendrils curling inward like charred paper before they vanished into the depths. The hum fractured, stuttered, then began to collapse.

For a heartbeat, Olivia's chest lifted with fragile relief. It had worked.

But victory was never clean.

Her eyes flicked to his face. Drops of seawater clung to his hair, catching the lantern light like beads of glass. He did not look at her, only at the wound, as though nothing else in the world mattered but setting it right.

For the first time since the tank had shattered, Olivia felt the pounding of her heart slow. The hum was still there, deep in her bones, but Alex's presence softened its edge.

A half-soaked notebook waited, pages curling where ink had bled into the grain. The scrawled words glimmered faintly in the lantern glow, as though the sea itself had stained them.

When Alex bound her wrist with practiced efficiency, Olivia found her gaze wandering from his hands to his face. He worked with the same patience she'd seen him use untangling snarled nets on the docks, calm, methodical, as if no task was too ruined to salvage. His fingers, calloused and strong, were surprisingly gentle as they wrapped the cloth snug around her arm.

She caught herself studying the lines of concentration carved into his brow, the steady set of his jaw. The lamplight traced the faint scar that ran from his cheekbone to the edge of his jawline, a mark she'd noticed before but never asked about. In the storm's flickering shadows, it looked less like an old wound and more like a badge, a quiet testament to battles endured.

For a moment, the ruined lab, the shards of glass glittering like ice, the endless hum vibrating in her bones, all of it receded. There was only Alex, close enough that the damp scent of salt and rain clung to him, steady enough that her trembling stilled.

"Thank you," she whispered, the words fragile, almost drowned by the storm raging against the windows.

Alex's eyes flicked to hers, only briefly, but the weight of that glance pinned her breath in her chest. He didn't linger. His focus returned to the knot he tied, pulling it firm but not cruel, "You can't keep doing this alone," he said, voice low, almost rough. The words weren't a scolding, they were too quiet, too weighted with something else. Concern. Resolve.

Olivia's throat tightened. She wanted to protest, to insist she could handle it, that this was her burden, her creation, her mistake to mend. But the way he said it left no room for deflection. He wasn't asking. He was stating a truth.

The hum swelled faintly, pressing against the walls. The tanks flickered with ghostly light, but for that small stretch of time, Olivia's fear was less sharp. Alex's presence steadied the edges of it, as if the storm outside could rage, the sea could rise, the jellyfish could pulse their strange signals, and still, here at her side, he anchored her.

When he left a few minutes later, lantern glow trailing into the storm-muted night, Olivia sat frozen, staring at the bandage. His careful knotting held firm, neat and certain in a way her own trembling hands could not have managed. Yet the glow still bled faintly through, pulsing with its unnatural rhythm, muted now, softened by his care, but no less present.

The lab groaned around her, beams settling after the storm, water dripping steadily into the pools on the floor. Each drop sounded louder in the silence Alex left behind, as if the world had shrunk back to its rhythm of damage and waiting.

At her desk, she pulled her notebook from the pile of damp papers. Half the pages were ruined, ink leached into gray fogs, words lost forever. But a handful remained dry enough to take her hand. She turned to one and forced the pencil across the page, each letter a battle against her fatigue:

Subject breached containment.

Injury sustained.

Persistent luminescence.

Correlation between wound and resonance of specimens.

Possible permanent alteration.

The graphite dragged unevenly; her hand shook, her wrist ached. When she stopped, the words swam before her eyes, dissolving into meaningless shapes. A brittle sound escaped her throat, too sharp to be a sob, too hollow to be laughter. She clamped a hand over her mouth, but it echoed anyway in the ruined lab.

Permanent alteration.

She had wanted resilience for the ocean. She had wanted to save reefs, to make the seas stronger, to give something back to the only world she'd ever truly loved. Instead, she had rewritten herself, her body carrying the proof of her meddling like a brand.

Her gaze shifted to the darkened window. Outside, the storm had passed, leaving only the sea's restless churn below. But the glass reflected her own face more than the view. And for a heartbeat, she did not recognize it.

Her skin was too pale, her eyes ringed with shadows, no, with light, faint halos that shimmered in tune with her heartbeat. Her shoulder, beneath the damp fabric, glowed steadily, the bandage a thin veil that could not contain it. The reflection shimmered, as though something beneath the surface pressed forward, waiting to emerge.

She flinched back, turning away sharply, breath ragged. Her pulse roared in her ears. For a moment she clutched the edge of the workbench just to steady herself. The hum lingered low in her bones, reminding her that she was no longer only observing her specimens.

She was becoming one of them.

She crouched before the hearth, the ruined pages heavy in her arms. Ink smeared her palms black, mingling with the blood from shallow cuts, until her hands looked like they belonged to someone else entirely. The paper sagged with damp, fragile as skin, threatening to tear apart before she even reached the fire.

One match. That was all it would take. One hiss of sulfur, one spark, and the work that had consumed her sleepless nights, that had cost her blood and perhaps her humanity, would be ash.

Her fingers found the matchbox on the mantle. She turned it over in her hand, the rasp of wood against cardboard loud in the silence. The creatures hummed louder behind her, the tanks vibrating softly, as though they too knew what she was about to decide.

Olivia's hand shook. She imagined the flames catching, imagined the relief of watching the words curl into smoke and vanish. No one could blame her then. No one could point to her notes and say this was the blueprint of their ruin. No one would ever know what she had done.

But the thought curdled almost as quickly as it came. If she burned it all, she burned the only map through this nightmare.

Every line she'd written was knowledge, dangerous, damning, incomplete, but knowledge nonetheless. And without it, Haven's Reach would face the tide blind.

She closed her eyes, the matchbox pressed to her forehead. The faint glow of her wrist lit the damp pages in her lap, a ghostly lantern, reminding her that she carried the evidence whether she wanted to or not.

Her grandfather's voice echoed in memory, worn with age but steady:

"The sea doesn't yield to fire or fury. It yields to understanding."

Her hand lowered. She placed the matchbox back on the mantle with deliberate care. Then, with her uninjured hand, she drew a single dry page from the wreckage, set it on the workbench, and picked up her pencil.

Her wrist burned, the hum pressed close, and still she wrote, her letters jagged but defiant:

Containment impossible.

They are more than experiment.

They are aware.

The pencil snapped in her grip, but she didn't stop staring at the words. For the first time since the glass shattered, she knew: there was no turning back.

Her whisper vanished into the air, devoured by the thrum of the tanks. The jellyfish drifted close, their tendrils brushing the glass as though straining to hear her vow. Their glow flickered, brighter for a breath, then dimmed again, like an acknowledgment, or a warning.

Olivia pressed her aching wrist against her chest, feeling the faint tattoo of light pulsing beneath her bandage. It beat in time with the creatures, a tether she could not sever. The matchbox still lay where it had fallen, its dull tin catching the glow like a half-buried relic in sand. Her gaze lingered on it. Burning everything would have been simple. Clean. But nothing about this was simple anymore. She had chosen the harder path, the only path she could live with.

She slid the folder shut, her hand still trembling. The paper crackled like brittle ice, fragile, but it was all she had: a record of what the town would never believe, and what she could not deny.

Lightning split the horizon, briefly painting her reflection in the windowpane: pale face, hollow eyes, and that faint, unearthly glow seeping through her sleeve. Not just a scientist. Not just a woman.

Something caught between.

Her jaw tightened. "I will fix this," she said again, louder this time, as though daring the sea itself to hear her.

The hum receded into a low, steady drone, almost like a heartbeat.

And for the first time that night, Olivia realized she no longer knew whether it was hers or theirs.

Twelve

A FRAGILE HOPE

By the time they reached shore, dawn had begun to gray the horizon. The storm-scattered clouds glowed faintly with the first light, their bruised purple giving way to the palest silver. The tide rolled in soft and steady; a lullaby compared to the violence of the night before.

For the first time in weeks, Haven's Reach lay quiet

No hum shivering in the air.

No flicker of ghostly glow threading through the waves.

The silence was not peace exactly, but absence, like a fever breaking, leaving behind a body still trembling in its recovery.

Olivia climbed from the boat with her grandfather's steady hand guiding her. Her legs trembled beneath her, each step heavy as though the ocean had tied weights to her ankles. Her injured arm throbbed with its stubborn light, pulsing faintly beneath the bandages, but the air itself felt lighter than it had the night before.

They dragged the boat higher up the shore, the sand resisting them, the tide whispering at their heels. Each scrape of the hull against stone seemed to echo louder than it should, as though the world itself was listening. Neither spoke. The only sound was the faint rattle of the serum vials in their crate, a reminder of how narrow their victory had been.

Olivia lowered the crate gently onto the damp sand. She crouched beside it; palms pressed against the salt-wet wood. For a long moment she didn't move, simply breathing in time with the soft lap of the waves. Foam curled around her boots, climbing and retreating, climbing and retreating. Ordinary surf. Nothing more.

No shimmer of phosphorescent blue.

No hum prickling through her bones.

No living light thrumming beneath the tide.

For a heartbeat, she almost let herself believe it. The sea seemed… itself again. Just water. Just tide. Just breath.

Her grandfather straightened slowly, joints creaking as he leaned on the oar. The movement was deliberate, measured, as though every bone and tendon remembered storms older than she had ever seen. He studied her face with the same care he once gave to tide charts and star maps, reading the exhaustion carved into her features, the hollow beneath her eyes, the faint blue shimmer still bleeding stubbornly through the bandage at her sleeve.

The lantern flame in his hand flickered, paling against the brightening horizon. Its glow seemed fragile beside the slow spread of dawn, a human-made light already giving way to something greater. He did not miss the way she hunched against herself, as though trying to keep the ocean's pulse from escaping through her skin.

"You see, child," he said at last, his voice softened by fatigue but tempered with pride, "the sea doesn't ask for mastery. It never has. It asks for respect. Tonight, you gave it what it was owed." His words rolled like the tide itself, inevitable, steady, impossible to ignore.

Olivia swallowed hard. Her throat ached with the weight of unshed tears, heavy as stones she could not lay down. "And yet it still took something from me," she whispered. She flexed her glowing hand, the marks pulsing faintly beneath the wrappings, the light slipping between threads like a secret too wild to keep. "It won't let me forget."

Her grandfather stepped closer, the sea wind tugging at his coat. He rested his weathered hand gently on her shoulder, warm and immovable as the cliffs themselves.

"Scars remind us where we've been," he said. His tone carried no rebuke, only truth born of long years. "They don't have to decide where we go."

The words sank into her like balm, though they could not erase the weight of her guilt. She gazed out over the horizon, where the sky bled scarlet into the restless sea, each wave capped with gold as the first light touched it. Somewhere beyond that line, beneath miles of water, she knew the jellyfish had retreated. Diminished, yes. But not destroyed. Never destroyed.

Her chest tightened as she thought of the fishermen who had abandoned their nets, their livelihoods shaken loose by the glow that haunted the waves. She pictured families huddled behind shuttered doors, whispering prayers against a hum that crawled through the walls at night. And she thought of her own hands, steady, brilliant, reckless, that had spliced nature until it twisted into something unbound.

Her voice was quiet when it came, spoken almost to the waves themselves: "I'll spend the rest of my life trying to make this right."

Her grandfather's hand tightened, a firm squeeze that carried both comfort and command. His gaze did not waver from the horizon, but his voice came low and certain. "Then that will be your redemption."

The sound of boots crunching against wet sand pulled her from the horizon. Olivia turned and found Alex walking up the beach, his dark hair damp with salt spray, his shirt clinging to broad shoulders from the mist of the sea. He had carried the oars the last stretch alone, muscles taut with the effort, silent as ever, his presence steady as stone.

He didn't speak again. He simply lowered the oars beside the boat, the wood sinking into the damp sand, then straightened with a sharp breath. His gaze flicked to her arm, the faint glow bleeding through the bandage, and lingered a heartbeat too long before he looked away, jaw tightening as if he'd seen more than she wanted to admit.

Something passed between them in the silence. Not pity. Not judgment. Something heavier, steadier, an unspoken recognition that went deeper than words. He had seen her falter in the boat. He had seen her clutch the crate with trembling hands; her skin lit like a lantern in the dark. He had watched her bleed light into the night. And still, he was here. Not to scold, not to flee, but to stand near enough that she felt, for a moment, less burdened by the weight of the sea alone.

The tide hissed over the sand and back again, filling the silence that stretched between them. Olivia's chest tightened.

She wanted to thank him, but the words caught in her throat, tangled with everything she could not yet name.

Her grandfather's cane struck gently against the sand as he shifted his weight, watching them both. His lantern flame wavered in the morning wind, and a shadow of a smile touched his weathered face, quick, knowing, and gone before Olivia could see it. But Alex caught it. His shoulders stiffened, his eyes dropping back to the wet sand, as though bracing against a truth left unspoken.

Olivia, still caught between guilt and awe, noticed none of it. She only rubbed at her arm, the glow pulsing faintly beneath her fingers, and turned her gaze back to the horizon.

Her grandfather started toward the village, lantern swaying with each step, its light bobbing like a second star against the pale wash of dawn. His figure grew smaller as he disappeared into the shadows of the dunes.

Olivia lingered, unable to make her feet follow just yet. The tide slid around her boots, cold but strangely gentle, tugging at her as though asking her to stay.

Alex lingered too. He stood a few paces away, his broad frame outlined against the gray horizon, the crate at his feet. The silence between them stretched, filled only by the hiss of waves retreating and returning.

"Thank you," she murmured at last, her voice frayed, nearly lost to the surf. She wasn't sure if she meant it for hauling the oars, or for catching the crate, or simply for staying when others might have run.

Alex's eyes flicked to hers briefly, sharp as a knife in the half-light, then dropped away just as quickly. "Don't thank me," he said, his tone rough but quiet. "Just... don't stop fighting."

The words struck something deep inside her, not because of their sound, but because of the weight behind them, a man who rarely spoke, who guarded his thoughts like treasures, offering her one regardless.

Before she could answer, before she could tell him what the promise meant to her, he bent in one fluid motion, hefted the crate of serum onto his shoulder, and followed her grandfather up the beach. His steps were steady, each one firm, as if he carried not only the crate but some unspoken vow.

Olivia remained where she was, the tide brushing her ankles, the horizon bleeding into gold and silver as the first rays of sun climbed the sky. Her arm pulsed beneath its bandage, but for once the light did not feel like a curse. It felt like a reminder that she was still here.

The waves whispered against the sand, carrying no hum, no glow, only the fragile hush of a peace that might not last.

And for the first time in many nights, Olivia let herself hope.

From the dune path ahead, her grandfather glanced back once, lantern swinging, his eyes shifting between the girl who lingered by the water and the man who carried her burden without complaint. The faintest curve of a knowing smile touched his weathered face before he turned and disappeared into the waking village.

Thirteen

HAVEN'S REACH

AWAKENS

The next morning, Haven's Reach woke to silence. No humming tremors threaded through the night air. No ghostly blue glow stained the tide pools or flickered across cottage walls.

Only the familiar chorus of gulls circling the docks, their cries sharp and ordinary, their wings cutting through the pale light of dawn.

The harbor smelled of salt and fish again, not ozone and fear. Nets hung drying between poles, creaking softly in the breeze. Water lapped gently against the pilings, and for the first time in what felt like years instead of weeks, the sea looked like the sea.

Women leaned from their windows to shake out linens, and children darted barefoot across the sand, laughing as they chased the gulls into flight. A fisherman whistled while mending his line, the sound strangely bold in the fragile calm.

It was the kind of morning the town had once taken for granted. The kind of morning that now felt almost too quiet, as though the silence itself might shatter if anyone breathed too loudly.

Olivia stood at her window, arms wrapped tight across her chest, the bandages beneath her sleeve faintly warm against her skin.

Below, the fishermen were already gathering along the shore, their voices carrying up the cliffside in jagged bursts of disbelief and hope.

"It's gone, I tell you," One insisted, hauling a heavy net hand-over-hand, shoulders straining with the familiar rhythm. His words carried the relief of a man who desperately wanted to believe them.

"Calm for now," another muttered darkly, pausing to rub the sign of the cross over his chest. "Best not to test it too quickly. The sea remembers."

But when the first boat glided back into harbor, its nets sagging with a gleaming haul of fish, a cheer erupted from the dockside. The sound rolled across the square like a tide of its own, pulling others into its current. For the first time in weeks, the smell of salt and silver filled the morning air instead of dread.

Children shrieked with laughter, darting along the sand to chase crabs into their holes, their bare feet kicking up sprays of water. Women pushed open their shutters wide, letting the dawn in with a clatter of wood and iron, their faces soft with relief. Somewhere near the square, the old church bell tolled three times, its deep voice booming over the rooftops in gratitude, its echoes mingling with the cries of gulls.

To the villagers, it was a miracle. The kind their grandparents told in whispers by the fire, storms calmed, nets filled, danger passing like a shadow in the night. Already the whispers began to shift from fear into wonder, from muttered blame into cautious praise.

But at her window, Olivia's arms tightened against her chest. To her, the silence that returned was not the silence of safety. It was the silence of a sea holding its breath.

Olivia pressed her bandaged arm tighter to her chest. The linen was damp from sweat, but it did nothing to hide the faint, steady glow that pulsed beneath her skin. Each beat of light was a reminder, of what she had unleashed, of what she had barely contained, of the tide of consequences that no cloth could bind.

The door creaked softly against the morning hush. Her grandfather entered without a word, lantern extinguished now that the day had claimed the sky. In his hands he carried a chipped mug, steam curling from the surface of the tea like mist over tide pools. He set it carefully on the table beside her, then rested his weathered palm on her shoulder. The warmth of his touch cut through her shiver more than the tea ever could.

"Let them have this morning," he said quietly, his voice worn thin with exhaustion and threaded with something gentler. "They've earned it."

Outside, Haven's Reach rang with life again, the bell tolling, gulls crying, fishermen shouting their triumphs to anyone who would listen. The sounds of normalcy, fragile and bright, spilled through the window.

Olivia nodded, but her chest tightened until it hurt to breathe. The mug sat untouched, the steam fogging her notebook where it lay open but blank. The serum had worked. That was the truth. But deeper still was the other truth: the swarm was not gone, only driven back, and the sea had swallowed something new and unnatural in its depths.

Her grandfather's hand lingered, steady, anchoring. But the guilt coiled tighter, clawing at her ribs. Haven's Reach might believe in miracles, but she knew better. Miracles never came without a price. And this one had been hers to pay, hers to bear alone.

Down in the market square, the town rejoiced. Nets overflowed with writhing silver, their scales flashing like coins in the morning sun. Merchants raised their voices over one another, hawking fish with the kind of laughter that had been absent for too long. Children darted between baskets, their hands sneaking toward stray minnows, while the tavern owner promised to fry the first catch by noon.

Amid the noise, a mother clasped her grandfather's hand with both of hers, clutching it as though it were a lifeline. "Your prayers were answered," she said, her eyes wet with gratitude. Her little girl clung shyly to her skirts, glancing up at the elder with wide eyes that seemed to reflect the entire sea. "The storm has passed. The sea's forgiven us."

Olivia caught the words through the open shutters of her window above, her tea cooling in her hands. *Forgiven.* The word slid into her chest like a blade. They thought it was prayer, faith, divine mercy that had calmed the waters. Not science. Not the nights she had bled herself thin over equations and specimens. Not her reckless hands that had bent life until it broke.

She pressed her bandaged arm tighter against her ribs, the glow beneath its wrappings a secret the town would not have understood even if she'd shouted it from the square. They didn't want to understand. They wanted miracles, clean and merciful, without the stain of consequence.

Below, her grandfather's face was unreadable. His eyes, deep-set and shadowed by years of watching tides rise and fall, lifted briefly to meet her gaze at the window. He didn't correct the woman. He didn't speak of science, or serum, or the cost that still lingered in Olivia's flesh. He only gave the mother a tired, gentle smile and patted her hand softly, as if to anchor her joy.

And then he moved on, lantern swinging faintly at his side, leaving Olivia to swallow the bitterness alone.

Later, as the sun climbed higher, Olivia walked the narrow streets with her sleeves pulled down low. People greeted her absently; their attention fixed on baskets heavy with fish. For the first time in weeks, the air smelled of frying oil and brine, laughter carrying through open doorways.

At the pier, she paused. The water stretched endlessly before her, foam tracing ribbons across the surface. The sea looked ordinary again.

Innocent. Yet she couldn't shake the sensation that the calm was only skin-deep.

A voice behind her broke her reverie, low and steady. "You're up early."

Olivia turned. Alex was unloading crates from his boat, the muscles in his forearms taut as he lifted with effortless strength. His shirt clung damp to his back, dark hair plastered to his brow, but nothing in his movements showed strain.

He set the crate down with controlled precision, then straightened to his full height. His eyes, sharp beneath the fringe of wet hair, flicked to her arm before locking on her face. He didn't comment on the glow. He never did. He only held her gaze with a stillness that made her pulse quicken before he bent back to his work.

"You were out late," she said softly, more accusation than question.

Alex rolled a shoulder in a quiet shrug, the motion powerful, unbothered. "The sea doesn't wait for daylight," he said, voice low, edged with salt and gravel. "Neither can we."

There was no boast in the words, just the quiet certainty of a man shaped by tide and storm. His hands, rough from rope and salt, gripped the next crate as if it weighed nothing. She noticed how the other men on the pier instinctively gave him space, how even his silence carried an authority that needed no words.

Olivia lingered, though she told herself she should move on. Something in his steadiness rooted her there. She wanted to thank him, for steadying her on the boat, for shielding the vials when the lantern shattered, but the words lodged in her throat. Instead, she said, "They think the sea has forgiven them."

Alex lowered the crate with deliberate slowness and rested his forearms against it, the veins in his hands still taut from the weight. His gaze lifted to hers, dark and unflinching, framed by the morning light off the water. "The sea doesn't forgive," he said. "It waits." The words rolled through her, deeper than warning, heavy as tide.

A chill traced her spine. She glanced back at the horizon, its blue deceptively calm. Alex followed her gaze, and for a heartbeat the silence between them seemed to hum faintly, not sound, but memory.

Olivia drew her arm tighter against her chest. "It isn't over," she whispered.

Alex didn't argue. His jaw flexed, his eyes lingering on her a moment longer. Then, with the faintest nod, he gave the impression not of a man conceding, but of one who had always known.

By noon, the town square rang with laughter and song. Fishermen raised their mugs of ale high, foam spilling over as they clapped each other on the back. Merchants hawked fish with boisterous voices, and children dangled their feet from the pier, kicking at the spray as though the sea had never been a threat at all. To Haven's Reach, the nightmare had passed. The hum was forgotten, swallowed beneath the clamor of voices eager to believe in peace.

But at the cliffside, apart from the revelry, Olivia stood alone. The gulls wheeled overhead, their sharp cries slicing through the illusion of calm. Her sleeve hid the faint glow on her wrist, but it could not smother the truth burning beneath her skin.

The hum was still there. Not loud, not violent, but present. Faint as a memory, steady as a pulse. It vibrated at the edges of her awareness, patient as the tide.

She gazed out over the restless horizon, where the sea stretched wide and deceptively calm. The water gleamed under the sun, a mirror of innocence, but Olivia's chest tightened with the certainty that it was a mask. The ocean was not healed. It was waiting.

And deep down, she feared that when it exhaled again, when the hum returned in full force, Haven's Reach, drunk on hope and blinded by relief, would not be ready to face what rose from its depths.

Fourteen

FIELD TESTING BEGINS

B y late afternoon, word had spread through Haven's Reach like wildfire: the sea was calm again, and the Turners had a hand in it.

Fishermen gathered along the docks, their caps clutched in calloused hands, their weather-lined faces drawn tight with uncertainty. Nets dripped brine in loose coils, gulls wheeled above, and the smell of tarred rope, salt, and drying fish hung heavily in the air. For weeks, fear had strangled their voices into whispers; now, they spoke in tentative tones, words rising cautiously against the gentle lap of the tide.

"You've always known the tides better than us," one said to Olivia's grandfather, his voice pitched low as though afraid of being overheard by the waves themselves. His gaze kept flicking to the horizon, where the water lay deceptively calm, glassy in the late light. "Whatever you did last night, we'll help see it done again."

Another man cleared his throat, shifting his weight from foot to foot. His rough fingers twisted the brim of his cap until the fabric strained.

"If it keeps the waters safe for our nets," he said, his tone gruff but tinged with desperation, "you'll have no shortage of hands. We can't starve while waiting on miracles."

A murmur of agreement rippled through the crowd, thin, uneven, but carrying a kind of fragile hope. Eyes turned to Olivia and her grandfather as though they held not just the cure for the sea but the promise of supper on the table.

Olivia stood a little apart, her bandaged arm hidden beneath her sleeve, the crate of vials resting heavily at her feet. She let her grandfather speak for them both; his voice, steady and hardened by years at sea, carried a weight hers could not yet muster.

From the back of the crowd, Alex leaned against a piling, arms folded across his chest. He said nothing, but Olivia caught the way he scanned the fishermen's faces, the way his jaw tightened whenever the talk grew too close to accusation. He had been there on the water; he had seen what she'd risked. Though he never spoke up, his presence was a kind of shield, silent, solid, firm as the tide at her back.

Olivia shifted her weight, the boards of the dock groaning under her boots. Every instinct told her to step back, to vanish into the shadows of the net sheds, but her grandfather's steady presence anchored her in place. His words rolled over the crowd like the tide, calm but unyielding.

"It won't be simple," he said again, his voice gravel-soft, carrying without needing to rise. "You've seen its moods. This will take patience, steady as the tide. But if we work together, the waters might yet forgive."

A murmur of assent rippled through the fishermen, though doubt still tightened their faces. Some shuffled their caps in calloused hands; others stared at the water as if expecting it to flare with blue fire at any moment. Hunger pressed at them, and fear too, but hope began to flicker again like the first spark of a lantern wick.

Olivia kept her gaze on the crate at her feet. Its faint glow bled through the wood, betraying her work even as she tried to hide it. She wanted to speak, to tell them what the serum could and could not do, but she could almost hear the whispers her words would provoke. Turner's curse. The witch of the cliffs. So, she kept silent, biting down on the words until her jaw ached.

From the edge of the gathering, Alex shifted. He leaned off the piling, arms uncrossing as though the silence in the air demanded his weight. He didn't raise his voice. He didn't need to. His eyes swept over the men with the measured calm of someone who had weathered harsher seas than most of them would ever dare.

When one fisherman wavered, mumbling about cursed waters and nets gone thin, Alex's gaze fixed on him. A small shake of the head. A faint tightening of his jaw. Nothing more. Yet the man straightened, as if scolded by the sea itself, and said no more.

Olivia's heart tightened. He had said nothing, not to her, not to them, but his silence carried weight, a shield as real as the one her grandfather raised with his words. She wondered, just for a breath, if Alex had come for the town's sake at all. Or if he had come because he knew she would be standing here, too exposed, too easy a target for the villagers' doubt.

They set out in three boats as the sun dipped low, painting the horizon with streaks of fire and gold. Lanterns swung from the masts, their glow faint against the vast sky. The smell of tar and brine clung to the wood as the oars dipped and rose, the rhythm of men used to long nights at sea.

Olivia sat at the bow of her grandfather's boat, cradling a vial in both hands. The liquid shimmered faintly, catching the last rays of sun, as if eager to prove itself. She held it close, as though afraid the ocean might snatch it from her before she had the chance.

Her stomach tightened with every stroke of the oars. Though the hum was absent, she still felt it in her bones, a phantom echo that pulsed in her wrist, in her shoulder, in her memory. The water looked calm, but it felt expectant, like a stage waiting for its actors.

Alex sat just behind her; shoulders bent to the oars. His strokes were measured, powerful, each pull steadying the boat even when the others wavered. He said nothing, but she felt the weight of his presence like a second anchor. Once, when the boat lurched sideways in the wake of a companion vessel, his knee brushed her back as he leaned forward to steady the line. She didn't turn, but the touch lingered, a quiet reminder that she wasn't alone at the bow.

Her grandfather guided them toward the cove where the jellyfish had swarmed thickest. His weathered hands gripped the tiller as if carved into it. "Ready?" he asked, voice quiet but firm.

Olivia nodded, her throat too tight for words. She lifted the vial, its glow painting her fingers pale blue. Alex's voice came low behind her, just one word, but weighted with steel: "Steady." One by one, they began. She smashed the first vial against the water, glass shattering, serum spilling into a luminous cloud that bloomed outward like liquid dawn. The fishermen followed, their nervous motions gradually giving way to rhythm. Vials arced through the air, breaking on the tide, each burst of light adding to the ripple of radiance that spread across the cove. Ripples of faint luminescence mingled with the sunset until sea and sky burned together in shades of fire and blue.

For long moments, nothing stirred. The silence was sharp enough to cut. The men shifted uneasily in their boats, muttering, waiting for something to break. Olivia gripped the gunwale so hard her knuckles whitened. Behind her, Alex shifted too, but not in impatience. His hand brushed the edge of her cloak, steadying it where the wind had tugged it free. She didn't move. Neither did he.

Then the glow below the surface began to shift.

Faint shadows wavered, then receded. The last traces of unnatural light flickered and died, snuffed out like candles in the wind. One fisherman gasped as he pulled his net, and it bulged with silver fish, alive and writhing, scales flashing like coins. Another boat let out a cheer as their traps came up full. The sound carried over the water, rough-edged and jubilant.

Olivia's heart pounded. For the first time since the glass had shattered in her lab, she felt a fragile spark of hope kindle inside her chest. Perhaps, just perhaps, this could work.

She turned slightly, just enough to see Alex in her periphery. His dark hair caught the last light of the sun, his face unreadable but his jaw firm. He hadn't cheered with the others. He only kept rowing, steady as the tide, his gaze flicking once to her bandaged wrist before returning to the horizon. That silence, that constancy, steadied her more than the cheers ever could.

By the time they returned, the village was waiting. Lanterns blazed along the docks, their flames swaying in the evening breeze. Children ran barefoot across the planks, their shrieks carrying over the tide as they darted between the men unloading baskets heavy with fish. Nets glistened, silver bodies writhing in abundance, and women wept as they threw their arms around their husbands, relief and hunger mingling in their voices.

The church bell tolled, its peals brighter than before, booming across the rooftops and echoing against the cliffs like laughter. Someone shouted, "You've done it! The sea is ours again!" and the words rippled through the crowd like a spark catching dry kindling. Cheers rose, rough and jubilant, filling the square until the air itself seemed to vibrate with them.

Olivia stood near the gangway, her cloak damp and heavy around her shoulders, the bandage beneath throbbing with its faint, stubborn glow. She forced herself to smile as villagers clapped her grandfather's back, their eyes shining with gratitude. But inside, the words stung like nettles. *The sea is ours again.* They didn't understand. The sea had never been theirs to claim. It never would be.

Still, she let their relief wash over her, if only for a heartbeat. After weeks of dread and silence, the joy of her people felt like sunlight breaking through stormclouds, fragile and fleeting though it might be.

On the edge of the square, Alex lingered apart. He hoisted a heavy crate of fish onto his shoulder as though it weighed nothing, his shirt clinging damp against his broad frame, his movements as steady and deliberate as the tide itself. He didn't join the cheering. He didn't lift his voice in thanks or clap another man's back. He only looked at her, a fleeting glance, eyes dark under the lamplight, and in that flicker, she caught what the crowd could not. No smile. No easy relief. Only something harder, heavier. Understanding, perhaps. Or warning.

Her chest tightened. She wanted, absurdly, to cross the square, to close the space between them and ask what lingered in his silence. But before she could take a step, Alex turned back toward the boats, his shoulders squared beneath the weight of the crate. His silence carried its own authority, louder than any cheer.

The villagers believed the sea had been tamed. Olivia knew better.

And judging by the look Alex had given her, so did he.

Later that night, when the villagers had dispersed and the square lay empty save for the creak of rigging and the murmur of the tide, Olivia and her grandfather stored the empty vials in the lab. The faint shimmer of residue clung to the glass like a warning, refusing to dim no matter how carefully they wiped the surfaces clean.

Her grandfather paused by the shelf, lantern in hand. The flame guttered low, catching the lines of his face. He met her gaze, his voice a gravelled whisper.

"Enjoy their joy," he said. "But keep your eyes open. The sea tests us, always."

Olivia nodded, though her chest felt cold. The cheers of the villagers still echoed faintly in her memory, but they no longer warmed her. She could still feel the hum, faint as an echo in her bones. She wrapped her cloak tighter and stepped outside, her boots crunching softly on the damp earth.

The horizon stretched endless and dark. The air smelled of salt and iron. She thought she saw, far beyond the lantern glow, a single pulse of light rippling in the deep. Quick. Subtle. Like a heartbeat.

She blinked, and it was gone.

Behind her, her grandfather came to stand in the doorway, his cane planted firmly in the earth. His lantern threw a halo around him, but his gaze drifted toward the docks where a lone figure lingered even at this hour. Alex, carrying the last of the nets to dry, his shoulders broad and shadowed against the moonlit sea.

Her grandfather's eyes twinkled faintly despite the weariness in them. "That one doesn't cheer loud like the rest," he murmured, more to himself than to her. "But he watches. Always watching."

Olivia frowned faintly, not catching the meaning, her eyes still drawn to the horizon. "He doesn't trust me. None of them do."

A quiet chuckle rumbled from her grandfather's chest. "Trust and care wear the same face, child. Some are just too stubborn to name it."

Olivia turned, brow furrowed. But her grandfather only lifted the lantern higher, letting its light fall back toward the path. "Come. The night's long, and tomorrow will be longer still."

She followed him, though as she glanced once more toward the dock, she saw Alex straighten from his work, his head lifting briefly toward the cliff. Their eyes didn't meet in the dark, but the weight of his presence followed her all the same.

And as she climbed back to the glasshouse, the faint shimmer beneath her bandage pulsed once, matching a rhythm she could not name.

Fifteen

THE COST REVEALED

At first, the days that followed felt charmed. The docks bustled again, baskets brimming with fish, nets heavy with silver scales. The tavern filled each night with voices raised in song, laughter spilling into the narrow streets until the sound seemed to thread through the very timbers of Haven's Reach. Children played tag between barrels stacked high with salted catch, their bare feet slapping against the cobbles, their giggles ringing like bells. For the first time in weeks, the village slept soundly, shutters flung wide to let in the ocean breeze without fear of the hum pressing through the walls.

The townsfolk whispered prayers of gratitude, some to the sea, some to the church, some to both. Wives clasped their husbands' weathered hands at the market, saying the tide had turned. Old Brant lifted his mug in the tavern and declared the danger past. "We're free," they said, voices light with relief, and the words spread like birdsong from door to door.

Olivia wanted to share in their joy. She wanted to believe the worst had passed, to let herself be carried on the current of relief. Yet when she looked at their smiles, she saw only fragile glass, ready to crack at the faintest tremor. Each morning, when she walked the cliffs to check the tide pools, unease grew at the edges of her mind like a shadow at her back.

The pools looked ordinary enough, barnacles crusting the rocks, crabs scuttling sideways, seaweed tugging gently with the tide. But now and again, she caught it: a patch of coral bone-white when it should have been living red, a wisp of glow lingering beneath the foam before fading. Once, she found a fish washed up on the rocks, its belly torn, veins faintly luminescent even in death.

She took notes, scribbled until her fingers cramped, but the words did nothing to ease her dread. The town's laughter filled the air behind her, but the sea gave its own answer in silence too deep to ignore.

THE FIRST SIGN: CORAL

It began with the coral. Where the serum had spread during their night on the boats, the reef showed its wounds first. Olivia crouched by the water's edge at low tide, the morning mist curling around her ankles. She reached for a cluster of coral she had known since childhood, once vivid with life, red branches teeming with darting fish and tiny snails. Now it was bone-white, as if years of sun had bleached it in mere days.

Her gloved fingers brushed it gently, unwilling to believe. The branches gave way instantly, brittle as old parchment. They collapsed to powder beneath her touch, dissolving into dust that shimmered faintly before vanishing on the tide like smoke. The sound it made was quieter than breaking glass, yet it cut sharper than she expected.

Her breath hitched, throat closing tight. She sat back on her heels, notebook balanced against her knee, and wrote furiously, her script sharp and angry:

Serum effective against specimens, but collateral damage present. Marine flora destabilized. Long-term effects unknown.

She underlined *unknown* three times until the page nearly tore. The ink smudged under her trembling hand.

The tide rushed in and curled back around her boots, foam hissing against the leather as though mocking her. She stared at the place where the coral had been, her mind replaying the night's triumph, the villagers' cheers. For every cheer, a fracture. For every prayer of gratitude, a cost no one else yet saw.

When she rose, the sea spray clung to her skirts, heavy as guilt.

THE SECOND SIGN: THE CATCH

Two days later, the pounding on her door startled Olivia from her notes. She opened it to find one of the harbor boys, no more than twelve, freckles scattered across his nose, chest heaving from the run up the cliff path. He didn't waste a word, only grabbed her sleeve with both hands and tugged hard.

"Come quick," he gasped. "You've got to see."

Her pulse quickened. She pulled on her gloves and followed him down the twisting path to the docks.

There, a knot of fishermen had gathered, their bodies forming a wall around a single wicker basket. Salt crusted their coats, nets hung limp over their shoulders, but none moved to sell or clean their catch. Their faces were carved with unease, their voices low and frayed.

Olivia pushed through. One of the men glanced at her, hesitation flashing across his weathered features, but he shifted aside all the same.

At first glance, the fish inside looked ordinary. Silver scales flashed in the morning light, wet and gleaming. Relief almost surged in her chest, until one of the men upended the basket.

The fish spilled across the planks, slapping weakly. And then she saw it.

Their bodies shimmered. Not with the sheen of oil or the glisten of life, but with something unnatural. Veins of pale blue pulsed beneath their skin, faint light even in death. Each pulse echoed faintly, like embers buried in ash, refusing to extinguish.

"They're tainted," someone muttered.

Another spat hard into the cracks of the boards, eyes darting to the horizon.

"Not natural."

A third crossed himself quickly, lips moving in prayer.

Olivia crouched despite the warning in her gut. She reached with gloved fingers and pressed gently to the gills of one fish. Its body twitched, a faint spasm, a dying shiver, as though something still lingered, tethered just beyond sight.

Her stomach lurched. She yanked her hand back, bile burning her throat.

The men's murmurs rose. "Can't eat that."

"Throw it back to the sea."

"Sea don't want it either," someone answered grimly.

Olivia stood, her face pale, throat dry. The crowd parted, letting her pass without a word.

Later, in the solitude of her lab, she forced her hand to write, though her script wavered and smeared across the page:

Serum absorbed into fauna. Unintended transference. Potential contamination of food supply.

She sat staring at the words long after the ink dried, the glow of the fish still burned into her vision.

THE THIRD SIGN: THE CHILD

The final warning struck closer to home. A fisherman's daughter, no older than eight, had wandered with her friends to the tide pools. Hours later, she returned crying, clutching her arm. Her mother carried her to Olivia in terror.

"She only touched the water," the woman pleaded. "She only touched it!"

Olivia lit a lamp, forcing her hands to steady as she examined the child's arm. A faint rash spread across her skin, glowing in the dim light like starlight beneath the flesh. Olivia's own bandaged wound pulsed faintly in reply.

The girl whimpered when Olivia pressed gently around the marks. Olivia whispered assurances, though her throat was tight. "She'll be all right. It should fade."

The words sounded thin even to her.

When the mother left, Olivia leaned against her desk, shaking. Her grandfather found her there, staring at the glow beneath her bandages as if it were a brand.

"It's the cost," he said quietly, laying a weathered hand on hers.

"There shouldn't be a cost," Olivia snapped, her voice breaking. "I meant to heal, not poison! What if the whole town,"

"The sea never gives without taking," he interrupted gently. "Balance always asks its price."

Her tears blurred the ink on her notes. "Then I've doomed them."

"No," he said firmly, tilting her chin toward him. "Not doomed. Warned. You know more now than you did before. That knowledge is your duty."

But when she returned to her journal, the words swam. Every line felt like a confession, every scratch of her pencil an admission of guilt.

ALEX'S UNEASE

That evening, as the tavern swelled with music and mugs clashed against oak tables, Olivia lingered at the edge of the square. Lanterns swung in the salt-stiff breeze, laughter carried down the alleys, and for a fleeting heartbeat she almost let herself be drawn into the warmth of it. The relief was infectious, pressing in on all sides like tidewater rising against the rocks.

But then her eyes caught on a solitary figure at the far end of the wharf.

Alex stood apart, boots planted wide, shoulders stiff beneath his salt-damp shirt. His silhouette cut stark against the wavering lantern glow, steady and immovable as the pilings he leaned on. The sea stretched out before him, black and endless, and he stared at it as though waiting for it to reveal its secrets.

Olivia hesitated, pulse quickening. She remembered the way his arms had steadied her in the boat, the quiet command in his voice when the lantern shattered, the weight of the crate balanced effortlessly on his shoulder. A man who said little, but spoke volumes in every gesture. Now his jaw worked in the shifting light, tense, as though chewing on words he couldn't bring himself to give voice to.

She gathered her cloak around her and stepped toward him, the music from the tavern fading with each pace. When she reached the edge of the dock, her voice came softer than she intended.

"You're not celebrating."

His eyes flicked to hers, dark, unreadable, then returned to the water. "Hard to drink to a cure," he said, low and rough, "when the cure looks worse than the sickness."

The words cut sharper than she expected. She bristled, drawing her cloak tighter against the wind. "You think I've ruined us."

Alex didn't answer right away. Silence stretched between them, filled only by the creak of ropes and the lap of waves against the pilings. He wasn't cruel, she knew that. He was deliberate, weighing every word as if each carried the heft of stone.

Finally, he said, "I think you're the only one looking past the laughter. That… worries me."

The softness in his tone cracked through her defenses. He wasn't condemning her. He was afraid. Afraid for her. Afraid for them all.

Her gaze fell to the planks beneath their feet, her voice scarcely more than a breath. "Then we share that worry."

For a long moment, they stood in silence, side by side. The sea exhaled against the rocks, lanterns swung, the faint strains of song drifted across the square. Alex shifted slightly, his hand brushing the weathered piling near hers. Close enough she felt the heat of him in the cool night air, a quiet tether that steadied her as surely as the oar in her grandfather's grip.

She didn't dare move closer. He didn't pull away. That was enough.

When she finally turned back toward the cliffs, the tavern song swelled again behind her, laughter spilling into the night. She glanced over her shoulder once, just once, before climbing the path. Alex was still there, dark against the horizon, eyes fixed on the restless sea as if waiting for it to breathe again.

At the top of the path, a lantern flickered in her grandfather's hand where he stood waiting. His gaze followed hers down to the solitary figure on the wharf, and though he said nothing, a faint chuckle rumbled low in his chest.

Above, the town celebrated.

Laughter spilled from the tavern doors, mugs clashed in clumsy toasts, and the church bell tolled as though it had chased away every shadow. Children darted through the square with sticky fingers and salt-tangled hair, their joy loud enough to drown the memory of fear. Haven's Reach clung to the illusion of peace with desperate hands.

Below, the reefs crumbled.

What had once been living gardens of coral now withered bone-white, brittle branches snapping like old chalk in the shifting tide. Fish slid between them with wary, unnatural light glimmering beneath their scales. The sea carried on in silence, but its wounds spread quietly, layer by layer, like ink bleeding through paper.

And in Olivia's bones, the hum returned, faint but steady.

It wasn't in the air, nor in the water lapping the cliff base, but inside her, as if her own marrow vibrated with the memory of the swarm. Every pulse whispered the same truth, patient and merciless: not finished.

Olivia pressed her glowing wrist to her chest, her breath shallow, her eyes on the dark horizon. The town above slept easier. The sea below did not.

Sixteen

FIRST SIGNS OF SPREAD

The evening air was heavy with salt when Olivia climbed the cliffs above Haven's Reach. She had walked that path since childhood, tracing the narrow grooves worn by fishermen's boots, but tonight each step dragged as though the earth itself resisted her climb. The cries of gulls had quieted with the sunset, leaving only the hush of waves rolling against the rocks.

The wind tugged at her cloak, carrying the mingled scents of tar, brine, and smoke from the tavern fires below. Lanterns glowed faintly in the village, dots of warmth scattered against the growing dark, but from this height they looked fragile, fireflies trembling at the edge of an endless sea.

Her breath came shallow, chest tight as she pressed on. She knew every turn of the path, every jutting stone where she and her mother once paused to catch the light of fireflies, every tide pool where her grandfather had pointed out starfish and whispered old tales of the sea's moods. But tonight, those memories offered no comfort. The weight of her bandaged wrist pulled at her side like an anchor, the faint glow beneath the linen throbbing in time with her heartbeat.

At the crest, she paused. The horizon stretched wide before her, water black as obsidian. For a moment, the ocean seemed calm, only the rolling tide, only the breath of the deep. But Olivia felt it still, hidden beneath the hush: the hum. Low, patient, faint as memory, but there all the same, waiting.

Below, the town glowed warm with lantern light. Laughter spilled from the tavern, punctuated by the scrape of fiddles and the thud of boots on worn floorboards. For the villagers, the return of fish to the nets and quiet nights free of humming had brought relief so complete it bordered on denial. Children darted through the square playing tag, their shadows long in the lamplight. The smell of frying oil and ale filled the air.

Snatches of song rose and fell, rough-edged but joyous. A fisherman raised his mug in toast, and the crowd answered with cheers that rolled out into the streets. Old Brant bellowed a story from the docks, his voice booming above the rest, while wives hushed their little ones with sticky hands and promises of honeyed cakes. Even the church bell, struck in celebration, rang not as a warning but as a song of reprieve.

From the cliffs, Olivia watched, the sound carrying up to her like echoes from another life. For them, the danger had passed. For her, the glow beneath her bandages pulsed like a lie. The joy in the square below felt fragile, built on the silence of things left unspoken, on the illusion that the sea had been tamed. She hugged her arms tighter around herself, as though bracing against a cold no lantern could drive away.

But Olivia carried the truth in her chest like stones in her pockets. The glowing fish. The bleached coral. The rash on the fisherman's daughter's arm. Signs of balance breaking, not restoring. Each memory pressed heavier against her ribs with every step up the cliff path, until she thought she might shatter beneath the weight.

She reached the cliff's edge and stood still, the sea wind clawing at her cloak, her breath quickening in the night air. The horizon lay calm, deceptively calm, the water stretched smooth as black glass, reflecting a thin sliver of moon. From that height it looked ordinary, harmless, like the sea of her childhood where she had once danced barefoot in the tide pools, laughing at the way minnows darted around her ankles. For a heartbeat, she let herself believe it again: that the serum had worked, that Haven's Reach could return to its old rhythm of tides and nets, of gulls and bells, of prayers whispered without fear.

Then she saw it.

A faint pulse. Far out in the deeps. Blue-white. There and gone, like the flicker of distant lightning trapped beneath the waves. Her breath snagged in her throat. She blinked, willing it to vanish as an illusion of tired eyes. But it came again. And again. Another pulse, then another, until the horizon was stippled with them. Dozens. Scores. A constellation shifting beneath the waves, cold stars in a restless sea.

Her pulse hammered. The sound of her own heartbeat filled her ears until it wasn't her heartbeat at all but the hum. Low. Patient. So, faint at first it might have been mistaken for the breath of the wind across stone. But Olivia knew better. This was not air or storm or rock. This was resonance, ancient and insistent, thrumming through her bones as though the sea itself whispered in her blood.

"They've retreated," she breathed, clutching the jagged stone at her side until it bit into her palm. "But they haven't gone."

Behind her, footsteps crunched against gravel. She stiffened, turning only when the glow of her grandfather's lantern spilled across the cliffside. He moved slowly, his weight leaning on the oar he carried as a staff, his silver hair damp with sea air. The flame lit half his face, leaving the other in shadow, the lines around his eyes etched deeper than she remembered.

For a long while, he said nothing. He simply stood beside her, shoulders squared against the wind, and followed her gaze into the restless dark. Together they listened to the faint hum threading through the night air, a sound that felt less like hearing and more like remembering.

"You feel it too," she whispered, not daring to raise her voice.

He nodded slowly, his jaw tightening as though against pain. "The tide draws back," he murmured, his voice a stone carried on the wind,

"Only to return stronger."

The words slid into her chest like ice. Haven's Reach had always trusted the rhythm of the sea, ebb and flow, give and take, balance and return. But this was something else. This was not balance. Not nature. It was a waiting silence, coiled and hungry.

Far below, the sound of wood groaning split the hush. Olivia flinched, her gaze snapping downward just in time to see a fisherman's boat limping into the harbor. Its mast leaned like a broken bone, sails ragged, nets dragging from the sides in shredded ribbons. The hull listed heavily, every creak a protest against the tide.

Lanterns along the pier flared to life one by one as the town stirred. Families rushed to meet the vessel, their shouts echoing up the cliffs. Olivia's grandfather shaded his eyes against the glow, his jaw tightening. From their height, they could see the men stumbling onto the docks, pale, shaking, their movements sluggish as if the sea itself had drained them dry.

Even at that distance, their voices carried, torn ragged by fear.

"Lights in the deep," one gasped, his words tumbling over themselves.

"Moving faster than current, faster than any current ought to run." Another man gripped the railing as though it still anchored him to life. His hands shook uncontrollably. "Something touched the hull. Not fish. Not weed. Tendrils. I swear it."

The third fell to his knees, his shoulders heaving, every muscle taut as though the memory clawed at his skin. His voice rasped, hoarse and broken. "The sea… it spoke. It whispered in my head."

The words curled up the cliffside like smoke, lodging in Olivia's chest.

She pressed her hand against the bandages at her arm, feeling the glow pulse beneath, faint but steady, an echo of the hum that vibrated through her bones.

Her breath caught. The serum had bought them time, yes. But not peace. Not safety. The jellyfish hadn't been defeated, only scattered. They had regrouped in the black water beyond the harbor; their pulses scattered like cold stars gathering for a storm.

Her grandfather's hand came down heavy on her shoulder, the weight of it grounding her as much as it warned her. His grip was steady, his lantern glow flickering against the deepening dark, but his voice was grim, carved from stone and sorrow.

"This is only the beginning, child."

Olivia swallowed hard, her eyes fixed on the broken boat below. The lanterns on the dock burned weakly against the encroaching dark, casting more shadow than light. The hum beneath her skin deepened, throbbing in painful sympathy with the sea. She clenched her glowing wrist against her chest, but the pulse only grew stronger.

Her whisper came out thin, but it carried truth like a blade. "Then Haven's Reach isn't safe. None of us are."

A QUIET WITNESS

Movement stirred along the slope below, subtle but enough to pull Olivia's eyes from the sea. A figure leaned against a piling where the cliff path wound toward the docks. Tall. Broad-shouldered. Lantern light caught in his damp hair and traced the salt-streaked lines of his shirt.

Alex Shepherd.

He hadn't joined the rush to the wounded fishermen, nor had he gone to the tavern where voices now rose in alarm. Instead, he stood apart as he always did, watchful, silent, his presence less a comfort than a long shadow stretched before dawn.

Olivia's chest tightened, her heart giving a treacherous flutter. She told herself it was only nerves, only exhaustion after a night without sleep. But her eyes lingered, drawn to him as if pulled by some unseen tide.

Alex bent, coiling a length of rope with the same calm precision he brought to every task. Each motion was deliberate, efficient, nothing wasted. Yet even in that quiet labor, she saw the flicker of unease shadow his features when the horizon pulsed again, faint and blue-white. His jaw set hard. He did not glance toward the tavern or the villagers clustering on the docks. His gaze stayed fixed on the black horizon, steady, as if he alone could hold it back.

From beside her, her grandfather followed her gaze. The lantern light etched his profile in gold, his expression unreadable, but Olivia caught the faintest tug at the corner of his mouth, a subtle knowing glimmer. He let it pass without comment.

Heat rose to her cheeks. Olivia tore her eyes away, forcing her focus back to the endless dark sea. Still, the weight of Alex's silent vigil pressed against her like the hum itself, not loud, not insistent, but impossible to ignore.

THE PULL OF THE SEA

The hum deepened, low and relentless, a vibration more felt than heard, as though the ocean itself held its breath beneath the stars. Olivia gripped the jagged edge of the cliff until her knuckles blanched white, the stone biting into her palms. Far out, where sky and water bled into one, the glow rose again. Swelled. Faded. Returned. Each pulse steady as a heartbeat, cold as a signal.

Her thoughts raced in rhythm with it, questions she could not silence.

Had the serum truly driven them deeper?

Or had it forced them to change?

Had she scattered the fire instead of smothering it, leaving sparks to spread across waters too vast to contain?

Her arm answered in kind, the faint light beneath her bandage flaring with each throb, as though her flesh itself echoed the sea's pulse. She longed to tear the wrappings away, to see if the rhythm in her veins matched the distant flicker on the horizon. But fear rooted her hand.

Some truths were worse to witness than suspected.

She turned, seeking refuge in the ordinary. From the cliff she could see Haven's Reach spread below, warm with lanternlight. Children still laughed in the square, their voices bright and unbroken. Fiddle music tumbled out of the tavern, careless and merry, each note a denial of the dread that had gripped them only nights ago. The smell of frying fish drifted on the breeze, homey, familiar. To the village, it was proof that the sea had relented. That life was theirs again.

But Olivia knew better.

Her gaze pulled back to the horizon, where the pulses throbbed like cold stars in a sea that refused to rest. Her heart matched them, her scar ached with them, and she could not shake the certainty rising like a tide inside her.

The fight had not ended.

The sea had only paused, waiting.

And she, unwilling heir to its whispers, would have to face what came next.

Seventeen

FRACTURES IN HAVEN'S

REACH

M orning broke clean and bright, the kind of sunshine that seemed almost cruel in its simplicity. The harbor glittered as though polished overnight, waves lapping gently at the pilings, their rhythm steady and calm. Ropes hung to dry along the docks, steaming in the golden light, their fibers soft with warmth. Gulls wheeled overhead, carving white slashes across a spotless sky, their cries sharp and ordinary, too ordinary. Even the church bell tolled its measured notes as though keeping time with a world that had not trembled in the dark.

But beneath the brightness, Haven's Reach carried the hush of a place pretending to be safe. The laughter of children sounded a shade too high, too forced. Fishermen's hands lingered a fraction too long on knots and nets, their eyes darting toward the horizon when they thought no one watched. It was a morning painted over a night of shadows, clean on the surface, but still breathing unease in the spaces between.

A hush spread like a ripple through the crowd. Even the gulls seemed to grow quiet, their cries distant above the square. Olivia felt the weight of eyes settling on her, more than she could count, each one pricking her skin sharper than the cold morning air.

She lifted her chin, though her heart thudded unevenly. "We used a compound to drive the jellies back," she said, choosing her words carefully, as though each were a fragile glass bead. "It was measured. Controlled. It is being monitored."

"Monitored?" Old Brant barked a laugh, but it was too loud, too sharp, a man's bravado stretched thin. "Monitored doesn't gut fish or feed children."

A boy perched on a crate whispered, "It's glowing," his eyes wide as saucers. His mother yanked him down, but not fast enough to stop the whisper from spreading.

Kellan folded his arms, cloth dangling from one hand. "You say controlled, Olivia. But look at this." He tapped the herring on his board, the faint blue shimmer pulsing once more as though answering his touch. "Does that look controlled to you?"

Olivia opened her mouth, closed it again. She wanted to say yes, to insist this was manageable, to promise the town safety, but the words stuck in her throat.

"It's Turner's work," someone muttered. "Always was."

Her grandfather stepped forward then, broad shoulders filling the narrow space between her and the murmuring crowd. He set his onion basket down with deliberate care. "Enough," he said, voice calm but iron-strong. "She's my blood, and she's trying to mend what none of you dared touch."

The crowd shifted uneasily, but the whisper had already seeded itself, curling through the air like smoke. Turner's curse. Turner's touch.

From the edge of the square, Olivia caught the faintest glimpse of Alex leaning against a piling, arms folded across his chest, his dark eyes on the crowd, and then, briefly, on her. He said nothing. He didn't need to. His steady presence was a counterweight, silent but heavy enough to matter.

But even with her grandfather's defense, even with Alex's quiet watchfulness, Olivia felt the tide turning against her.

The air thickened with unease. The market's ordinary clamor, the clink of scales, the slap of fish on boards, the bargaining of wives and sailors, had fractured into a silence broken only by the gulls overhead. Their shrill cries seemed mocking, sharp enough to scrape the skin.

Olivia shifted her weight, wishing she could fold into her cloak, wishing she could make herself as small as the apology she had given. But her grandfather's steady presence at her side anchored her. His hand hovered near the basket of onions, not clenched, not reaching, simply there, a quiet shield.

Tamsin's daughter peeked again from her mother's skirts, the glow in her wrist brightening as her small hand flexed. It caught the attention of others nearby. Eyes darted. Mouths tightened. A fisherman's wife crossed herself; another spat to the ground.

"Working," Olivia repeated softly, though the word felt fragile as paper now. Her gaze drifted to the fish glimmering on Kellan's board, the shimmer within their flesh echoing the pulse under her own skin. She pulled her sleeve tighter.

"Closing your stall won't change it," Brant grumbled, dumping fish into another basket, but even his gruffness sounded forced, hollow.

Kellan's shutters slammed down with a heavy finality. The sound cracked across the market like a gavel. The crowd twitched with it, shifting back from Olivia as though the air around her carried contagion.

From the edge of the square, Alex's figure caught her eye again, still and silent against the piling, arms crossed. His jaw was hard, unreadable, but his gaze moved over the crowd before settling briefly, just briefly, on her. A quiet warning, and a quiet reassurance at once.

Olivia swallowed. She had come to the market for food, for a shred of normalcy, but found only judgment coiled tight around every glance. She could feel the whisper waiting, gathering weight, just a breath away from breaking into a word she dreaded: *Witch.*

She forced herself to stand tall. If she flinched now, the story would write itself without her.

The crate's thud echoed like a warning, cutting sharper than any word. Heads turned. Conversations stilled. Even Brant's hands froze mid-motion, fish slipping wetly back into the barrel.

Alex straightened slowly, wiping one palm on his damp shirt. He didn't lift his chin, didn't raise his voice. He didn't need to. The sheer solidity of him, the way his shoulders squared as though bracing against the whole town, pulled the square taut.

"None of you like being judged for the nets you cast or the storms you fail to read," he said at last, voice low but carrying. "Don't do it to her."

It wasn't a defense so much as a line drawn in the sand. A line no one wanted to cross.

The murmur that had been swelling faltered. Someone coughed into their fist. Tamsin bent to hush her daughter. Brant grumbled, but quieter this time, his eyes flicking to Alex before returning to his own stall. Even Kellan, hard-eyed Kellan, looked away first.

Olivia stood motionless, her arm clutched close, the heat of humiliation burning under her skin. But beneath it, something else stirred: a fragile, unspoken relief. Alex hadn't looked at her once while speaking, and somehow that made it stronger. He wasn't performing for her sake.

He was simply refusing to let the crowd tip into cruelty.

Her grandfather, who had kept carefully silent through the flare of rumor, finally spoke, his voice pitched mild but meaningful: "Best tend to your catch before the sun spoils it."

The crowd, restless and chastened, began to dissolve, retreating into the safer noise of bargaining and bartering. Yet the air still buzzed with what had been seen: the boy's hand, the bandage, the glow. That would not be forgotten.

Alex bent to lift the crate again. As he passed Olivia, his sleeve brushed hers. He didn't slow. He didn't speak. But the steadiness of his stride, the deliberate weight of his boots on the boards, left the faintest impression: she was not standing alone.

"Catch is mixed," Alex said, his tone as flat as the tide. "Some clean. Some not. You can see it if you've got eyes and sense enough to use 'em." He tipped the crate with one hand. Half a dozen herring slid into Kellan's basin, their scales flashing in the morning light. The crowd leaned forward as if drawn by a string. Some fish lay plain silver, as ordinary as supper. But one or two carried the faint traceries of light beneath their skin, veins glimmering like frosted threads.

Alex pointed once. "You sell these." His finger brushed the clean. He shifted it to the others. "You salt or toss these. You don't scream at each other in the gaps."

Brant scowled, barrel-chested and loud by habit. "Since when do you run the market, Shepherd?"

Alex lifted his gaze at last. No flare of anger, no raised voice. Only the weight of someone who had weathered storms larger than Brant's bluster. "Since nets started coming up strange," he said. No heat. Just a fact. "And since you all decided to talk louder than you look."

The words landed heavy, cutting through the restless mutters. A gull cried overhead, wings flashing white against the sky, and for a moment it was the only sound.

A few men muttered again, softer now. One rubbed his palms against his thighs as though trying to rid himself of an unseen chill. It wasn't that Alex was beloved, he had never sought their approval. It was that he was immovable, a piling driven deep into the seabed. You could lash against him, curse him, even resent him, but you didn't shift him. You argued with Alex Shepherd the way you argued with the weather: briefly, and without expectation.

Olivia stood very still. She didn't thank him. She didn't dare. The words pressed against her lips, desperate to escape, but she swallowed them down. She kept her gaze on the warped planks beneath her boots, her breath even, her cloak tight around her. If she looked up, she was afraid she would find something in his face, a softness, a steadiness, that the town would see and twist into rumor.

Her grandfather stepped into the silence her restraint left behind. Broad-shouldered, weathered as the pier itself, he lifted his chin so the morning light caught the gray in his hair. "We keep each other safe," he said, voice low but carrying. "Like we always have. We listen for the sea's temper. We take heed. We don't turn on our own."

The authority in his words settled on the crowd like a hand pressing down. A few heads dipped. Tamsin hushed her child and stepped back into the safety of the group.

Kellan's jaw worked, but he said nothing more. Even Brant, blustering by instinct, looked away first.

The square loosened, shoulders relaxing by inches. Fish changed hands again, coins clinked, the market staggered back into motion. But the shape of it had shifted. The whispers would not vanish. The light in the fish would not be forgotten. And Olivia knew, deep in her bones, that something in Haven's Reach had cracked.

On the far side of the square, the bell in the church tower tolled noon. A woman crossed herself. A man spat into the gutter. Brant sold another basket. Kellan closed his shutters and slid the iron latch with a heavy finality that made Olivia flinch.

Her grandfather touched the small of her back, the way he had when she was little and too close to a steep stair. "Home," he said. "You're not a spectacle, child."

They walked the lane toward the cliff path. Even with her hood up and her head down, she felt the eyes, curious, resentful, afraid, like a fine rain soaking through wool. Two girls whispered and then pretended they hadn't. A fisherman she'd known since childhood nodded to her grandfather and not to her. She accepted it like penance.

They were nearly at the alley that cut to the steps when a voice carried after them, clear as a bell:

"Witch!"

It wasn't shouted like an accusation, or hissed like a curse, just said, the way you might say *gull* or *rope* or *net*. A fact trying itself out on the tongue.

Her grandfather stopped. He didn't turn. His hand tightened around the oar he used as a walking stick until the knuckles went pale. "Keep walking," he said, very mildly. And because Olivia suddenly wanted to turn and see whose mouth had made that word, she obeyed him.

But Alex had heard it too.

He stood at the edge of the market, rope coil slung over one shoulder, salt still drying in his dark hair. His jaw clenched hard enough to crack stone. The word had hit him like a thrown stone, and for a moment the air around him seemed to thrum with the weight of what he didn't say.

Witch. They spat it so easily, as though knowledge were a crime. She'd gone inland, to schools none of them could imagine, and then she had come back, back to this village, back to the cliffs, back to the sea that would swallow her long before it swallowed them. She was the only one who had ever left and returned, the only one who had knowledge that stretched beyond the nets and the tide charts. And for that, they called her a name they didn't even understand.

His hands curled tighter around the rope, veins standing sharp against tanned skin. He could almost see himself crossing the square, could almost see what it would mean to shut the mouth that had spoken. But he didn't move. He didn't need to. His silence was sharper than a shout, his gaze iron-edged against stone. The fishermen near him shifted uncomfortably, as if they'd suddenly felt the pull of the tide against their ankles.

He kept his eyes on Olivia's retreating form, her hood drawn low, her shoulders squared in stubborn dignity, and thought, not for the first time, that she was the smartest soul in Haven's Reach, and perhaps the loneliest.

If the sea ever did rise against them, Alex knew, it wouldn't be prayers or nets that stood between the village and the deep. It would be her.

And the rest of them were too small-minded to see it.

At the mouth of the alley, a shadow detached from the wall and fell into step a few paces behind. Boots quiet. A coil of line over one shoulder. Alex, head down, not speaking. Just there. The word *witch* had made a shape in the air that stuck to skin; his presence cut it thinner.

They reached the steps. Wind rose to meet them, clean and briny, lifting the edge of Olivia's cloak. Below, the tide scoured stone, patient and tireless. At the first landing, she stopped and made herself breathe.

"Let me speak at the meeting tonight," she said, surprising herself.

Her grandfather shook his head. "There's no meeting."

"There will be." The certainty came from a place she didn't recognize, a hard, bright line under her ribs. "Kellan will call one, or Brant, or Tamsin will ask the women to gather. If I don't show them, I'm not hiding, they'll make my absence into a story. Let me speak." He studied her for a long count, lantern flame trembling in the wind. "Truth won't change fear," he said at last.

"No," Olivia admitted. "But I have to try."

A soft scuff on stone. Alex had paused two steps below, as though he'd meant to pass and had changed his mind. He didn't look up when he spoke. "If you speak, speak plain. Bring something to do, not just something to say." He shifted the coil of line higher on his shoulder, the movement taut with contained energy. "People mend better when their hands are busy."

Olivia blinked, caught off guard by the practicality of it, and by the rough kindness behind the words.

Her grandfather's mouth twitched, almost a smile. His gaze slid from Olivia to Alex, then back again, sharp even in age. "Plain words," he said, as if repeating Alex's counsel. "And bring work for them to take home. The people trust hands before they trust heads."

"Plain words," Olivia echoed, though her eyes flicked briefly toward Alex before she could stop herself. He still wasn't looking at her, jaw set, watching the tide below like it might rise at any second.

Her grandfather turned up the steps, lantern bobbing. "Some counsel," he murmured under his breath, just loud enough for her to hear, "is worth more than it admits itself to be."

Olivia tucked the glow away, for the moment contained.

Olivia waited for the noise to crest, then took one step forward. The sound didn't so much fade as tilt toward her, like birds on a wire shifting their weight all at once.

Her cloak hem brushed the stone floor, damp from the rain that had followed them in. Every face seemed carved out of lanternlight and shadow, men she'd grown up knowing by name, women who had once pressed sweets into her hand as a child, now watching her as though she were a stranger.

"It's true," she said, her voice steadier than she felt. "The serum changes more than the swarm. It touches the fish. The reefs. Even those who come too near the tide pools." Her eyes flicked, unbidden, to Tamsin's daughter. The child stirred in sleep, her wrist faintly glowing through thin fabric. Olivia forced herself to look back at the crowd. "But it's not poison. It's balance shifting. And if we stop now, the sea will reclaim everything. You've seen what waits in the deep."

A murmur rose, uncertain, uneasy. Brant muttered some -thing about curses. Kellan pressed his lips thin. Tamsin's gaze burned.

From the side aisle, Alex shifted. Just enough to draw eyes without a word. His arms were crossed, broad shoulders braced against the stone, but the weight of his silence carried. He didn't look at Olivia, not directly, yet when Brant opened his mouth to sneer, Alex's jaw tightened, and Brant shut it again. Olivia's grandfather stepped forward then, lantern raised, his voice a gravelled anchor. "We don't speak of curses. We speak of choices.

The sea has always tested us. This time it gave us knowledge, and knowledge asks a price. My granddaughter has paid it already." His hand brushed his oar-staff, steady as tide. "Will you let her pay it alone?" The church went still, every breath loud in the hush. Olivia felt the heat of every stare, her bandaged wrist throbbing beneath her sleeve. She wanted to shrink, but instead she stood straighter, lifting her chin. Across the nave, Alex finally raised his eyes. Just for a heartbeat. Enough for her to catch the steady, storm-dark weight of them before he looked back to the crowd. He didn't speak. He didn't need to. The space he occupied spoke enough: he believed her.

And in Haven's Reach, belief could be louder than any argument.

"I should have come sooner," Olivia said. Her throat burned, but she forced the words out. "I should have said... more than I did." The church's stillness pressed down on her until she thought she might choke.

"I tried to make something that would hold the sea together when it was failing. I did it the wrong way. I hurt things I wanted to heal. I may have hurt you."

No one moved. A baby shifted in the back pew, its mother shushing quickly, as if even that sound might shatter the silence.

Olivia let the admission hang like an anchor dropped into deep water. Her voice cracked when she added, "I can't take it back. But I can make it better."

"How?" Kellan asked. No sharpness, no accusation. Just a man with calluses on his hands and fear in his eyes, desperate for something to hold.

"We're changing the mixture," she said. Her words stumbled, then steadied. "We can localize it. Shrink the spread so it doesn't touch the reefs. We can,"

"And the fish?" Brant broke in, folding his arms like a shield. "You going to dive down and pull the glow out scale by scale?"

"No," Olivia said. Her pulse thudded against her wrist, the glow hidden under linen. "But we can test each catch. We can mark what's safe and what's not. We can feed the town without guessing."

"And if there's nothing safe?" Tamsin's voice came, low and steady, her daughter's thin wrist glowing faintly where it peeped from beneath her skirt.

The harbormaster shifted, cleared his throat, and startled himself with the sound of his own voice. "Then we ask the boats to wait. A week. Two. Let the water settle."

"That's work gone," Brant barked, his words heavy with old hunger.

"It's bellies full later," Kellan countered. His eyes never left Olivia.

From the side aisle, Alex's voice cut through, low, rough, but steady, like the weight of tide against stone. "I'll show you how to rig bluecloth over the gutting tables. The glow shows up quicker under it."

Heads turned. Even Brant stilled. Alex didn't look at Olivia, or at them. He kept his gaze on the floorboards, as though the wood itself needed convincing. "I'll take first watch on the outer buoy. If the hum rises, we pull back in. No arguing in the dark."

The silence that followed wasn't fear this time, but consideration. Alex's words, simple and practical, landed like oars hitting water: direction, not noise. Olivia felt the knot in her chest loosen just a fraction.

Then, from the back, Mrs. Callahan's voice cracked through the hush like an oar across a hull. "It's not the fish, or even the glow," she said. Every head turned, because hers was the oldest voice in Haven's Reach, as weathered as the sea. She leaned forward on her cane, shawl slipping from her shoulders, eyes bright as stormlight. "It's the humming. You've all felt it. You know it's come and gone, and it's come again."

Her hand trembled as she lifted it, not toward Olivia, but beyond, through the stone walls, through the lantern smoke, toward the invisible line where the horizon met the sky. "You can call her names all you want. You can argue about pennies and scraps till you're blue. But the sea's got its own mind tonight." Her voice dropped, sharp as a gull's cry. "If you want to live, you listen."

The church held its breath. Lanterns hissed faintly. Olivia stood straighter, every eye still on her, but the word witch no longer clung so sharply.

The room rustled like wheat in the wind. Men shifted their boots, women adjusted shawls, children stirred in their mothers' arms. The harbormaster straightened, the decision settling on him like a cloak.

"Two-day halt," he said. His voice carried like a gavel on wood. "We test what's already caught. We don't sell what shines. We keep watch for the hum. And we don't throw stones at those trying to fix it."

The words hung in the nave, solid as stone. A few men muttered under their breath, but the fight had gone out of them. Brant's jaw worked, his mouth opening like he meant to argue, but then he shut it again, his mulish expression settling into something closer to reluctant consent. He gave a stiff nod, more to himself than anyone else. Tamsin's shoulders loosened by a hair, her daughter's small head tucked safely under her chin. Kellan exhaled so hard it sounded like he'd been holding his breath for a week, his arms falling from their defensive fold.

Alex hadn't moved. He leaned in the shadow of the aisle, arms still crossed, his expression unreadable. But when Brant's nod rippled through the room, Alex tilted his chin in the smallest motion, like a man marking a heading. The others saw it, Brant, Kellan, half the fishermen in the nave, and no one spoke further.

Then a sound found its way through the stone. Not the bell. Not the wind.

The hum.

Faint, almost imagined, it walked across the nave floor and climbed into bones.

Heads turned as if yanked by the same wire. Lantern flames quivered, tongues of fire whispering against the glass, thin as if the air itself were shying away.

It lasted only a heartbeat. Two. Then faded. Returned again, softer this time, barely there, like a memory or a promise.

Someone made a small sound in their throat; the kind people make when they don't want to cry. Someone else laughed too loudly, the brittle edge of fear wrapped in bravado.

The harbormaster clapped his cap against his palm.

"Watches and ropes and tides," he said, voice brisk, almost sharp, as if naming ordinary things could sweep the otherworldly away. "That's the work. Get to it."

People nodded too quickly, too eagerly. They gathered children, shuffled toward the doors, muttering about nets and storms and early mornings. But their eyes betrayed them. Every face tilted once more toward the black line of sea beyond the rooftops before they slipped into the lane.

When the last of them left, the nave went hollow. The heavy door thumped shut, and its echo wandered up into the rafters, then died.

Olivia didn't move. Neither did her grandfather.

And Alex, though half-hidden by shadow, stayed too, silent, jaw tight, as though waiting to see if the hum would return. His gaze flicked once toward her, sharp as a thrown hook,

before dropping back to the floor.

The three of them stood in the empty church, bound not by faith or miracle, but by the knowledge of what still waited in the sea.

Without being asked, Alex crossed the stone and set a small bundle on the front pew: a square of that dull blue sailcloth, a coil of twine, a lantern whose glass he'd smoked with soot. "For testing," he said. "You can take them." His gaze flicked to her bandage and away. "I'll walk behind you up the steps." A beat. "Just in case."

Her grandfather pretended to busy himself with his cap. "We'll manage," he said, but there was no bite in it. Only gratitude in disguise.

Between their palms, beneath her ruined bandage, the light flickered once and went still.

"Fractures heal," her grandfather said, eyes on the far wall. "Given time. Given care."

"And if we've cracked the bone of the town?" Olivia asked.

"Then we splint it and we keep watch," he said. "And we don't pretend we didn't hear the sea." Outside, the gulls screamed. The market shutters slammed. Somewhere on the docks, a rope snapped and men swore. The town shifted around its new fault line, settling into it like a body learning how to carry a changed weight.

On the steps, Olivia paused and looked back into the dim nave. The glass saints peered down with expressions no one could read. She wished, fiercely and stupidly, for them to blink and step down to tell her what to do.

She turned. Wind off the water hit her full in the face, salt, cold, honest. Alex fell in a pace behind, far enough that the town wouldn't make a tale of it, close enough that if she stumbled on the stair, he'd have her by the elbow before she struck stone.

She pulled her cloak tight and followed her grandfather into the thin noon light, steadying herself for the next tremor she knew was coming. And for a breath, a small one, the size of a match flame, she was grateful for the sound of a second set of footsteps keeping time with her own.

Eighteen

THE RETURN OF THE HUM

The first night passed in uneasy quiet. Haven's Reach slept, shutters latched, lanterns burning lower than usual, as though dim light could hide them from what stirred beyond the waves.

Olivia lay awake in her narrow bed, her arm glowing faintly beneath its bandage, the pulse of it steady and traitorous. She listened to the sea breathe through the cracks of the shutter, long, dragging sighs against the cliffs, too slow, too deliberate. Every creak of the house made her heart lurch. The roof timbers groaned like old men shifting in their sleep. A loose latch tapped and tapped against the window frame until she could no longer tell if it was wood or the faint echo of the hum returning.

She turned onto her side, pressing her arm against the mattress to dull its glow, but that only made the ache sharper, the light brighter in her mind's eye. Sleep never came. Instead, she counted heartbeats and gusts of wind, each one convincing her that the sound would swell, that the jellyfish would be there when she opened her eyes.

Down in the square, a dog barked once, sharp, startled, then went silent. Olivia held her breath. The silence after was worse than the sound. By the second night, the stillness had become unbearable.

The town had convinced itself that perhaps the worst had passed. Fishermen drank late in the tavern, laughter rising thin against the darkness, and children dared one another to run down to the tide pools where the glow had lingered weeks before. A fiddler sawed at the same tune twice as fast, as though speed could push fear away.

Olivia stood at her window again, unable to rest. The sea looked black and endless, the stars above clear as glass. Nothing stirred. Nothing threatened.

And then, without warning, the hum came back.

It began low, a vibration more felt than heard, shivering through the floorboards beneath her feet. The glass panes in her window trembled.

Her lamp flame quivered, its glow shrinking to a nervous flicker.

The hum deepened, stretching through the night air until the entire house seemed to breathe with it. The beams above her head groaned as if straining against the sound. Dust sifted from the rafters, sparkling briefly before settling on her desk.

Olivia pressed a hand to her wrist. The glow flared under the bandages in perfect rhythm with the hum, answering it, betraying her. Her throat closed. She stumbled back from the window as though the sea itself could see her there, standing in the faint light.

Down the cliff, she heard the first door slam open, then another. Voices rose, half-shouts, half-prayers, as lanterns flared across the harbor. Dogs barked, a panicked chorus cut short when the hum surged louder, swallowing their cries.

The sound was not outside anymore. It was in the walls, in her ribs, in the fragile glass of every lamp in Haven's Reach. It had returned not as a memory, not as a threat, but as a presence.

And it was stronger.

Somewhere down the lane, a dog began to howl. Doors slammed. Olivia clutched the sill with both hands, her glowing arm burning beneath its wrappings.

She could hear it clearly now, not just as sound but as resonance inside her bones, a steady thrum, rising and falling like the beating of some enormous heart. It felt purposeful, organized. Not the chaotic wail of a storm, but a signal.

She stumbled from the room and into the lane. Other doors had opened too; neighbors stood with lanterns clutched in white knuckles, their faces pale and fearful. Curtains twitched in upstairs windows. A child's cry cut sharp through the drone and was hushed instantly, swallowed into the night.

"It's louder," someone whispered. "Stronger."

The church bell swayed once in its tower, unstruck, the hum rattling its iron frame. Horses stamped and snorted in their stalls. The newborns in the cottages began to cry all at once, their wails twisting with the drone until it seemed the whole town was vibrating.

Olivia's grandfather appeared at her side; his face carved with lines of worry. He didn't speak. He didn't need to. His hand settled briefly at the small of her back, steadying her, before they both turned toward the cliffs, toward the black water rolling in.

They were not alone. Figures moved along the lane in the same direction: wives in shawls thrown on backward, men with boots unlaced, children carried like bundles. Lantern flames wavered under the pressure of the hum, making every face flicker between shadow and fear. The tide itself seemed to pull them outward, drawing Haven's Reach toward the cliffs like filings to a magnet.

At the corner by the sail loft, Alex Shepherd fell in with them without a word, a lantern cupped in his palm, its flame bowing to the invisible pressure. He glanced at Olivia once, the lantern's light cutting a bright nick across his cheekbone, then looked back to the sea as if he could anchor the town with his eyes alone.

"Keep to the path," he said, voice low and even, and the people nearest him obeyed as if the words were part of their feet. One boy tripped in the press; Alex steadied him with a firm hand, never breaking stride, his gaze fixed forward. The lane narrowed, and the hum swelled until the very stones underfoot seemed to pulse. Lanternlight spilled out across the cliff's edge, gilding the rocks in trembling gold. The tide pools were alive. From the highest ridge, Olivia saw them, pools shimmering like bowls of stars overturned on the rocks, glowing brighter with every wave.

Light lapped against stone in ragged arcs, sliding over wet rock and turning the black tide pools into mirrors of the night sky. Tendrils slithered and writhed beneath the surface, their pale bodies tracing patterns that dissolved before her eyes, pulsing in perfect rhythm with the hum. The sea was no longer just a backdrop; it was a living instrument, vibrating with the call.

The hum climbed a notch, a pressure change that made teeth ache and lantern glass whine. Even the cliff itself seemed to vibrate under their feet. The outer buoy lantern, far beyond the harbor mouth, flashed once in the distance as its glass shuddered in sympathy. The church bell answered with a nervous moan, its iron frame trembling without a hand upon its rope. Olivia pressed her palm to her chest. Her wound throbbed in time with the resonance. She gasped as the pain flared, the light beneath her bandage bleeding hotter with every note, as though something inside her were not resisting but answering the call.

"They're gathering," she whispered, the words pulled thin by the wind.

"Not just one. Not just here. The whole swarm."

Her grandfather's lantern shook in his hand, throwing erratic arcs across the stone. His lined face was set hard, but his eyes were wide and wet with the reflection of the glow. "Listen," he murmured, voice rough as weathered rope. "There's a count to it."

Olivia forced herself to hear past fear, past the roar of blood in her ears, past the shrieks of gulls retreating inland, and found the pattern. Thrum, pause, thrum-thrum, longer pause. Again. Again. Like a drumbeat being taught, patient and insistent. The hair on her arms lifted, her skin prickling as if the sound itself had hands.

Beside her, Alex raised one broad hand and tapped two knuckles against the wooden rail at the overlook, the sound small but steady. Thrum... pause... thrum-thrum... His lips barely moved. "Signal cadence," he said, almost to himself. "Like buoy calls. Or fog code." He tapped again, this time with more force, as if driving the meaning into the wood. "No storm speaks in numbers."

Olivia tore her gaze from the water to look at him, her breath catching at the calm, sharp way his profile cut against the glow. "You've heard this before?"

"Not this." He shook his head once, dark hair damp with spray, jaw tight. "But I've heard men try to talk to each other across weather. Short. Long. Repeat." His eyes tracked the dark runnels where water slid off stone and back into the sea. "This isn't a cry. It's a muster."

The word landed in her gut like a stone dropped into deep water, sending ripples through everything she thought she understood. Muster.

Not a call for help. A call to gather. To prepare.

Her grandfather closed his eyes, the lantern rattling faintly in his grasp. "Then the sea has found its army."

The hush was so complete that even the gulls refused to cry. Smoke from a few guttering lanterns drifted upward in thin, uncertain lines, as if afraid to stir the air too much. Mothers clutched children closer, not daring to soothe them. Men with ropes coiled at their belts stood as though turned to stone, every muscle strung tight, waiting for the next vibration.

Olivia's heart slammed against her ribs, too loud in her own ears. She still felt the cadence echoing in her bones, that final doubled pulse rolling through her like a warning that hadn't finished being spoken. Her glowing wrist throbbed in sympathy, light seeping faintly through her sleeve like a coal refusing to die.

Beside her, Alex kept his stance square, shoulders braced against the dark. He hadn't lowered his arm from when he'd warded the children back, and for a strange, unsteady moment Olivia thought he looked like part of the cliff itself, rooted, immovable. His eyes never left the black line of sea. "They didn't leave," he said quietly, voice low enough that it might have been meant for her alone. "They went quiet." Her grandfather let out a long breath through his nose, the sound more like a prayer than a release. "A silence that deep is never mercy," he muttered. "It's the sea drawing breath."

The villagers began to shift uneasily, voices rising in uncertain fragments: Was it finished? Should we go back? What was it? No one asked Olivia directly, but she felt the weight of their glances all the same, hot as coals against her skin.

She wanted to speak, to explain what she had counted, what she suspected, but the words lodged in her throat. She remembered Tamsin's child's glowing rash, the fish veined with blue light, the reef that had crumbled under her hand. What comfort could she give them now?

Her grandfather stepped forward instead, lantern lifted. "Back to your homes," he said, not harsh but firm enough that men obeyed without question. "Doors shut, lamps trimmed. The sea will tell us when it's ready to speak again." One by one, the villagers drifted back down the path, casting frightened glances over their shoulders. Olivia stayed at the ridge, unable to tear her eyes from the black horizon. Alex lingered, too, silent as always. His hand brushed the rail again, two knuckles tapping once against the wood as if testing the stillness, measuring whether the pattern would return.

It didn't. Not yet

And Haven's Reach exhaled, but no one truly breathed easy.

A fisherman finally broke the silence, his voice raw and frayed. "It's calling them."

Another whispered, hoarse as though afraid the words themselves might summon something closer. "From where?"

Olivia didn't answer. She couldn't. The truth was already coiled inside her chest like wire: the jellyfish were not only returning, they were speaking. Each pulse, each cadence, had carried intent. The hum wasn't noise anymore. It had become language.

The weight of that realization pressed down until the murmurs around her blurred, a sea of fear she couldn't wade through. Rosary beads clicked in Mrs. Callahan's hands like the ticking of a clock. The wind plucked at the shingles of the nearest house, making them chatter like teeth. Somewhere behind her, a child whimpered, shushed too sharply by a mother whose own voice trembled.

Olivia's knees gave a little as she stepped down from the ridge. Her grandfather's hand caught her elbow, steady and warm despite the night chill. He didn't speak; he didn't need to. His touch was enough to keep her from folding under the weight of what she carried.

Alex was there too, lantern dipping low to throw light across the path. He didn't look at her, but the flame caught in his jawline, sharp and unyielding, and for a moment the steady set of his shoulders grounded her too. His silence wasn't emptiness. It was ballast.

At the lower pool, she stopped. The water shimmered faintly, and there it was: a single ribbon of light drifting beneath the surface, thin and deliberate, like a vein glowing under skin. She crouched before she could think better of it, her breath fogging over the water's glow. Slowly, against every whisper of caution, she lifted her bandaged wrist just above the pool.

The scar beneath the wrappings answered. The faint blue brightened in sympathetic rhythm, pulsing in time with the light below. The connection made her stomach twist, nausea and dread tangled with something stranger, harder to name. "Don't."

The word was low, rough, startlingly close. She looked up. Alex had come to stand almost at her shoulder, his lantern throwing both their shadows long across the rocks. His voice wasn't sharp, but it carried weight, frayed at the edges not with anger but with fear he didn't know how to voice. "Not tonight," he said again, quieter now, like an oath.

She stood slowly, her legs stiff. The ribbon of light drifted on, indifferent, vanishing with the next push of tide. But the echo of its pull lingered in her bones.

Down the slope, Mrs. Callahan sank onto a rock, rocking as she crossed herself four times, lips moving in a prayer too old for anyone else to catch. Tamsin clutched her sleeping daughter to her chest, fingers locked so tight around the girl's wrist that her knuckles whitened. Kellan stood with his cap crushed in both hands, staring at the pools like a man who had watched his livelihood fracture and run out with the tide.

The harbormaster arrived at a jog, his breath pluming white in the night air, his cap clutched against his chest. "All hands, watch at the mouth," he said, not to the crowd but to Alex, as though some unspoken line of command had already been drawn. "No boats out. We rig lantern shades if it starts again." Alex nodded once, the motion clean, certain. "Bluecloth over the gutting tables at first light," he said back. "Glow shows quicker that way. We test the catch before anyone fillets."

The harbormaster bobbed his head, already turning to relay orders. It wasn't the clipped rhythm of command and obedience. It was older, simpler, like the passing of a coil of rope from one hand to another. No slip. No drop. Olivia watched, something tightening in her chest as the town, her town, gravitated toward Alex's voice, his steadiness, while her own words still felt like tinder waiting for a spark.

And beneath her bandage, her arm pulsed once more, as though the sea itself had given its answer., Olivia looked to the horizon where the buoy had flashed.

"There will be more," she said to no one, to everyone. "Maybe not tomorrow. But they're gathering."

"Then we gather too," her grandfather answered. "In daylight. With clear heads."

As people began to drift back toward their doors, Alex remained at the lip of the path, the lantern's soot-stained glass turning its light a dull amber. He stared seaward, jaw set, as if he could keep the ocean in place by squaring his shoulders at it. When Olivia passed him, he shifted just enough to speak without meeting her gaze.

"If it's a language," he said, "can you learn it?"

The question slid through her like cold water. Could she? Should she? "Language implies intention," she said. "And intention... implies strategy."

"Then we need one," he said. He lifted the lantern a fraction, its glow catching the edge of her sleeve where light bled through. "And we need you breathing to make it."

It was the closest thing to tenderness he had ever let loose in her direction. It landed like a promise and a warning both. She nodded, unable to trust her voice.

Back in the lane, the town creaked and settled as if it were a ship riding a swell. A shutter banged twice. A mother sang under her breath to a waking child. Somewhere the church bell swayed again, testing its own weight, then went still.

Olivia stood on her doorstep a long time, listening for the next tremor. The night seemed to hold its hands open, empty. Her arm throbbed in faint reply, a counter-beat she could not quiet. She pressed her palm flat against the doorframe, the wood cool under her skin, and tried to match her breathing to the honest rhythm of wind and wave.

Behind her, her grandfather set the bolt and laid his weathered journal on the table. "We'll map the intervals at first light," he said. "Old ways for old tides. New ways for new."

Olivia stared into the dark window until she could see in it the suggestion of her own face: tired, bandaged, haloed with a weak blue not even the wrappings could contain. Somewhere out beyond the buoy line, a constellation moved beneath the sea where no sky should be.

She whispered to her reflection, to Alex's question still hanging in her chest, to the sea that had begun to speak: "Then teach me."

Far below, the pools lay quiet. The stars above pretended indifference. Haven's Reach did what towns do in the pause between disasters, it slept badly and made plans. And out past the line where the bell could not carry, something in the deep answered, too soft for ears, just strong enough for bone.

Nineteen

THE FIRST ATTACK

The sea gave them no warning. At dawn, three boats set out from Haven's Reach, their sails snapping open to catch the pale wind. Dew still clung to the ropes, and the harbor smelled of tar and salt and hope. Men moved briskly, voices hushed, as though to raise them louder might wake the hum from slumber. Onshore, the villagers lined the docks. Mothers pressed shawls tighter around their shoulders, shading their eyes as the sails leaned into the breeze. Old men stood with caps in hand, lips moving in half-forgotten prayers. Children clapped and called, trying to stitch laughter into the air before fear could take it back.

The first boat bore Brant and his sons, their nets coiled neat, their faces sharp with determination. The second carried Kellan, whose scowl was as steady as the tiller in his hands, though his crew laughed too loudly, as if mocking their own nerves.

The third was Alex Shepherd's. He stood at the prow for a long moment before casting off, shoulders squared against the horizon. His crew was small, two men younger than him, one graybeard older, but no one doubted their steadiness. Alex's presence lent the boat its ballast as much as the stone in its keel. He had taken the outer berth without being asked, the place closest to the open sea, and no one contested it.

From the shore, Olivia watched. Her chest tightened as the sails filled and the boats leaned into the current. She told herself she was watching them all, but her eyes found Alex again and again: the sure set of his hands on the ropes, the way he turned his head once toward the cliffs as though marking them in his mind, then looked only forward.

For hours, the sea seemed kind. Nets dragged heavy, gulls wheeled, and the bright crack of laughter drifted faintly across the water, carried back on the breeze. By noon, the women of Haven's Reach laid frying pans ready and set out herbs, and the tavern rolled its casks into place. For the first time in many days, the air of Haven's Reach felt lighter, as if the village itself dared to believe in ordinary life again.

But Olivia could not shake the unease twisting inside her. The sea had given them no warning before. And she knew too well it would not need to again.

But by midday, only two boats returned.

They came crawling into the harbor like wounded animals, sails tattered, rigging frayed. Brant's vessel dragged low in the water, nets trailing loose and empty, while Kellan's listed hard to one side, a mast splintered at the top. The men aboard looked gaunt-eyed, hollow, their sunburned faces carved with fear instead of salt and wind.

The docks filled with bodies before the boats had even tied off. Wives and children pressed forward, voices rising in sharp, frantic waves. But no one asked the question aloud. They didn't need to. Every gaze went past the battered hulls, out to the gray line of the horizon, counting masts. One. Two. No third.

The absence bled into the crowd like ink into cloth. Shouts turned to whispers. A woman dropped her basket, fish spilling across the planks in a writhing silver flood, and no one stooped to help her. The bell in the church tower did not toll, but in the stillness, every creak of the ropes, every slap of the waves against the pilings, sounded like a funeral note. Olivia stood rooted among them, her hood drawn low, her hand pressed hard against her bandaged wrist where the glow had begun to stir. She searched the horizon until her eyes burned, desperate for a flash of sailcloth, a silhouette, a sign. The third boat, Alex's boat, was nowhere.

Beside her, her grandfather's jaw was tight as he leaned on his oar. "The sea keeps its own," he said softly, too low for anyone else to hear.

His voice was not comfort. It was warning.

And though no one spoke his name, Olivia felt it moving unspoken through the crowd like a ghost: Alex Shepherd.

When the third vessel appeared at last, staggering into the harbor with the tide, it was no ship at all, only a husk of timber. Its mast had snapped clean in two, its hull scarred with black streaks that pulsed faintly blue, as if the wood itself still smoldered with malice. Nets hung in ribbons. Ropes smoked where they had burned through, though no flame had touched them.

The crowd pressed forward in silence, faces drained of color. Olivia shouldered through the bodies, her grandfather's hand steady at her back, guiding her like he had when she was small. The smell hit her first, salt and copper and something acrid, unnatural.

"What happened?" she demanded, her voice cutting through the hush.

Two men survived. They clambered from the wreck, clothes plastered to their skin, salt-streaked and blackened in places as if they had been seared. One collapsed outright, his knees striking the planks with a crack. The other clutched the rail with blistered hands, eyes staring too wide, too bright.

"They rose," he rasped, and the words fell like stones into still water.

"God help us, they rose."

The crowd shivered. Someone crossed themselves. A baby whimpered.

"Tell us," Olivia urged, stepping closer, though her own pulse thundered. "What did you see?"

The fisherman's throat bobbed. His voice cracked, but the words came in fits and gasps.

"We heard the hum first. Louder than before. The sea shook with it. Then, then the water lit up, all around. Blue fire, everywhere." His eyes darted to the horizon, as though he still saw it burning there. "They weren't swimming anymore. They were… marching. Shoulder to shoulder. Their tendrils bound together. A moving wall."

The second man coughed up seawater, voice raw. "They struck the hull. Not once, but over and over. Like they knew. The wood split under us. The light burned straight through the planks. It wasn't chance." His hand shook violently as he pointed toward the broken mast. "It wasn't nature."

The dock groaned under the weight of silence. No gull cried. No rope creaked. Haven's Reach listened, and for the first time, the word 'attack' sat unspoken in every mind.

Her grandfather stepped forward, his presence filling the gap. "Back inside," he told the villagers, his voice steady as tide. "See to your doors. See to your children. We'll tend the wounded." But the people did not move. Fear had rooted them.

Then someone whispered it aloud: "They're hunting us."

The spell broke. Shouts erupted. Mothers pulled children close, fishermen spat curses, the harbormaster bellowed for everyone to clear the docks. The survivors were carried toward the infirmary, their skin faintly glowing where venom had kissed them.

Olivia caught her grandfather's gaze. His eyes were heavy with the knowledge they had both feared: the serum had not stopped the swarm.

It had bought them time, nothing more.

That night, the church bell tolled without pause, its iron throat groaning across the cliffs as a warning. Haven's Reach lay locked and dark: doors barred, nets abandoned, fires banked low. Even the tavern, the heart of the village, had shuttered its windows tight. The sound of its laughter, so recently reborn, was gone again. The only voice left was the bell, heavy and relentless, echoing like the heartbeat of a dying beast.

The sea breathed heavy in the distance, its rhythm less tide than threat. Now and then a wave struck stone with unusual force, as though something vast shifted just beneath the surface. The gulls had vanished hours ago, abandoning the

roofs and docks to silence.

In her laboratory, Olivia bent over her desk. Her notes sprawled in smudged columns of numbers and frantic sketches, lanternlight throwing long shadows across the ruined boards. Her eyes blurred with fatigue, but no rest came. Every few heartbeats, the bandaged arm at her side pulsed steadily, each throb synchronizing with the faint hum that never fully left the air.

She pressed her palm hard against the wound, as if sheer force might silence it, might stop the traitorous glow that answered the sea's call from beneath her own skin. Her teeth clenched. "They're adapting," she whispered into the quiet room, the words sounding brittle and foreign.

"Learning. Answering."

The hum seemed to swell in reply, faint but undeniable, like laughter just beyond hearing. "Child."

Her grandfather's voice carried from the doorway, low and grim. Lantern glow carved deep lines into his weathered face, shadows settling in hollows that made him look older than she had ever seen him. His hand trembled slightly on the frame, though his voice did not.

"The fever's back," he said. "And this time, it burns hotter."

Olivia closed her eyes, nails digging crescents into her palm. The hum pressed against her skull until she swore, she could feel her bones vibrating with it. In the silence between the bell's groans, she thought she could hear a cadence forming, a new rhythm, not random, not without meaning. The fever wasn't just in the sea. It was in her. And it was spreading.

Later, when the town had fallen into a restless half-sleep, Alex came to her door. He carried no lantern; he moved in shadow, shoulders broad enough to block the moonlight at her threshold.

"You shouldn't be out," she said, surprised at the rasp of her own voice.

He shrugged, though the motion looked stiff, the fabric of his shirt pulling against the fresh bandage on his ribs. "Can't sleep when the sea mutters like that." His gaze shifted past her to the scattered notes on her table, the faint glow bleeding from her sleeve. His mouth tightened. "You're worse than the rest of us. You look like you're carrying it inside you."

She almost denied it, then stopped. "Maybe I am." Something flickered in his eyes, worry, anger, something softer he wouldn't name. He stepped into the lab without asking, leaned on the workbench. The air between them seemed smaller suddenly, filled with the faint scent of brine and woodsmoke clinging to his coat. His movements betrayed a hidden limp; when he shifted his weight, his jaw clenched.

"You're not alone in this," he said at last. "Even if you keep trying to be." Her throat tightened. She busied herself with the bandage at her wrist, tugging it higher to cover the glow. His hand caught hers, rough, calloused, steady. He didn't hold on long. Just long enough for the contact to ground her, to say what words could not.

"You're still hurt," she said quietly, her eyes darting to the stain spreading through his shirt at the shoulder, the fresh edge of a scar across his forearm. "There's a guest room. You could stay here tonight. Rest, at least."

For a moment, he looked like he might accept. His shoulders eased, and something unspoken hovered between them. Then he shook his head, slow, firm. "The town would talk. And I've never slept well under roofs. I'll mend better where I can hear the sea."

Her chest ached with words she didn't dare speak. Instead, she nodded, fingers curling against the cool wood of the bench.

"Don't let the town see you break," he said, softer now. "They need you steadier than they need me." When the door closed behind him, she pressed her palm against the bench where his hand had brushed hers, as if to hold onto the warmth a little longer. The room felt emptier than before, though the hum still pulsed faintly through the walls.

Far out at sea, unseen by Haven's Reach, a wall of light pulsed across the waves. It stretched from horizon to horizon, not scattered shapes but a single living front, a curtain of cold fire. A thousand bells beat in unison, their glow swelling and falling like the breath of some vast creature. Tendrils twisted together like ropes, writhing and locking until they formed a lattice that moved as one.

The water itself seemed to bow beneath their passage. Waves rose and fell in rhythm, driven not by wind but by will. Schools of fish scattered before them in silver streaks, only to vanish into the dark where the glow did not reach.

And the hum rolled with them, steady as a drum. Deep, resonant, deliberate. A heartbeat that did not belong to any one creature but to the entire swarm. It carried through the deep and up into the black sky, echoing faintly across the currents like a war song.

They were not drifting anymore.

They were marching.

Twenty

A DESPERATE GAMBIT

The church bells had not stopped for three nights. Each toll carried a reminder across the cliffs: Haven's Reach was not safe, the sea was not theirs, and the glow had returned with vengeance.

Olivia's laboratory reeked of salt and burnt glass. The once-ordered shelves were a ruin of shattered vials and half-scorched parchment. Papers littered the workbench, stained with ink, seawater, and her own frantic hand. Some pages were smeared beyond legibility, her notes blurring into black swirls where ink had bled into salt. Others still bore neat calculations at the top but dissolved by the bottom into frantic lines: *More strength. More heat. More light.*

Vials glowed faintly in the lamplight. Some fizzed with unstable foam; others lay cracked in their trays, hissing faintly where the chemical burn had eaten into the wood beneath. Each failure seemed to drain the air thinner, until every breath she drew carried the acrid tang of ruin.

Her grandfather sat by the hearth, silent, watching her work. His hands rested on his cane, his face half-hidden in shadow. He had spoken little these last nights, but his presence had been constant, like the tide: unavoidable, patient, unyielding.

"You're trying to cure a fever by burning the body," he said finally.

Olivia didn't look up. She shook another vial, watching the liquid swirl into pale light. "I'm trying to stop it before it consumes us. The serum worked once. If I make it stronger, it can end this."

"Or end us sooner."

The words struck her like a lash. She slammed the vial down harder than she meant to. It rattled against the tray, glowing brighter, threatening to burst. "I can't do nothing. I won't."

He sighed, the sound weary and deep as the sea, but said nothing more.

At dawn, Haven's Reach gathered on the cliffs to watch. The bells had grown hoarse, their clangs dulled by salt air, but the people came anyway, drawn by fear and hope. Mothers clutched children, men twisted caps in their hands, and even the gulls wheeled silently overhead, as if they too were waiting.

Olivia and her grandfather pushed their small boat into the tide. The air was sharp with cold, the water black and heavy as iron. No other fishermen dared sail that morning. The harbor watched them leave, faces pale against the mist, whispers curling behind them like a shroud.

"She's going to end it."

"She's going to end us."

"She's the witch of the cliffs."

"She's the only one who's tried."

Olivia heard them all. None silenced the throb in her chest.

"You should have stayed ashore," her grandfather said as the oars creaked in his hands.

"I can't," she murmured. Her fingers clutched the crate at her feet.

Dozens of vials rattled within, brighter than any she had brewed before. They hummed faintly with their own inner light, as though alive. Even capped, they felt dangerous, like coiled serpents waiting to strike.

When they reached open water, she felt it first, the hum.

It rippled through the boat, rattling its planks, crawling into her bones. Blue sparks flickered beneath the surface, first one, then many, gathering, multiplying. Tendrils stirred below, twisting toward the light above.

"Now," she said, her voice shaking.

Her grandfather pulled the cork from the first vial. Together they hurled them into the sea, one after another, glass shattering, shards vanishing in clouds of liquid fire.

The water bloomed with white-blue radiance, blinding in its intensity.

The swarm convulsed. Bells pulsed violently. Tendrils lashed and recoiled from the glow, as though seared. The hum fractured into shrieks so sharp they splintered Olivia's skull. She staggered back, clutching the side of the boat, nearly vomiting from the vibration crawling up her spine.

For a moment, she thought it was working. The swarm thrashed, twisted, disintegrating into ribbons of light. She gripped the rail, breath ragged. "It's breaking them,"

Then she saw what followed.

The fish surfaced first, dozens, then hundreds, floating belly-up, their scales burning with an unnatural, horrifying brilliance. Their eyes bulged, glowing like lantern glass. Birds dove greedily, wings catching the glow, and dropped lifeless into the tide. Seaweed along the rocks curled and blackened, leaching light until nothing green remained. Even the barnacles on the hull cracked open, bleeding faint sparks before crumbling into dust.

Olivia's stomach twisted. "No… no, it wasn't meant to,"

Her grandfather seized her arm. His eyes burned with grief and fury.

"You've found a way to kill them," he said, his voice like splintering stone.

"And perhaps everything else with them."

The boat pitched as waves surged, the swarm dissolving into fragments that glowed even as they died, poisoning everything they touched. The air itself thickened, a luminous fog that seared Olivia's throat.

She doubled over, coughing, choking on light.

Her notes slipped from her lap, fluttering across the boards. She grabbed for them with trembling hands, her mind a storm: The serum worked. It worked. But the sea itself reels from the blow.

They rowed back in silence, the crate empty, the horizon dimming behind them. Haven's Reach watched their return with hollow eyes. Villagers lined the docks, silent as ghosts, watching the tide carry in floating corpses and weed turned to ash. No one cheered this time.

That night, Olivia sat at her desk, ink dripping onto the floor, the lantern guttering low. She wrote in jagged, uneven lines:

The serum can destroy them.

But destruction spreads.

The ocean will not forgive us.

Her arm glowed fiercely beneath its bandages, pulsing in time with the faint hum that still lingered beyond the walls. The swarm was not dead.

Merely angered.

She pressed the pen harder, carving the truth into paper she wished would burn itself. Her hands trembled. Her chest ached.

Then came the knock.

She startled, nearly upsetting the lamp. Her grandfather had already gone to bed, his steps slow on the upper floor. This knock was heavier. She opened the door and found Alex standing in the mist, shoulders hunched, his hair damp with spray.

"I heard what happened," he said simply.

Olivia shook her head, voice raw. "You heard what I did. Don't dress it as anything else."

Alex's gaze lingered on her sleeve, where faint blue pulsed through the linen. His jaw flexed, but he didn't look away. "You've carried this weight alone long enough. Don't think Haven's Reach hasn't noticed."

She wanted to snap, to tell him that all of Haven's Reach saw was a curse on two legs. But the words lodged in her throat. Alex's eyes, dark, unreadable, steady, held her in place.

"You came here, why?" she asked, almost a whisper.

"Because someone had to." His voice was low, but there was an edge to it. "And because you've bled for this more than anyone else."

Her chest tightened. She looked down at the glowing arm she hated and, for the first time in days, did not feel entirely monstrous under his gaze.

He shifted, clearing his throat, breaking the moment. "Get some rest," he said gruffly, turning as if the thought of staying any longer were dangerous for them both.

But Olivia caught the doorframe, her heart thundering. She wanted to call him back. She didn't. She only watched as his figure disappeared into the mist.

Her arm throbbed again, pulsing with the hum she could not silence.

For the first time that night, she didn't feel quite so alone in it.

Twenty-One

THE VOW

That night, the bell was silent at last. No one dared ring it, though the absence of sound felt heavier than its toll. Haven's Reach lay shrouded in darkness, the lamps surrendered their light, and hearth fires whispered low. Even the tavern was closed, its windows black for the first time in memory. The harbor stretched empty as a hollow ribcage, masts rocking in the still water like bones creaking in the tide.

From the cliffs, the sea spread black and endless, broken only by faint pulses of blue that pricked the horizon. They came and went irregularly, like the flashing of some monstrous heartbeat beyond the nets, beyond the prayers, and beyond the reach of anyone on shore.

Olivia sat at her desk, journal open, lamp guttering low. The little flame fought against the dark as stubbornly as she fought against despair. Her bandaged arm burned like a coal beneath its wrappings, the glow seeping through no matter how tightly she bound it, as though the swarm itself breathed beneath her skin.

She dipped her pen and wrote:

The serum works, but at a cost greater than we can bear.

The swarm adapts. Each time we strike, it grows stronger.

Haven's Reach cannot stand alone against this tide.

Her hand stilled. Ink pooled at the end of the last word, bleeding into the page. She listened.

The house was quiet, her grandfather dozing in the chair near the hearth, the rafters sighing with the weight of salt wind, but beneath that quiet ran another rhythm. Not in the walls, not in the sea outside, but in her own blood.

The hum.

Faint. Insistent. Patient.

Calling.

She closed her eyes against it, but closing them only made it worse.

The resonance tightened like a fist around her ribs.

A knock broke the spell.

Olivia startled, pen clattering against the desk. For a moment she thought she'd imagined it. No one knocked at this hour, not when fear kept the village shuttered. Then it came again, slow, deliberate, steady rather than urgent.

She rose, wrapping her cloak tighter, and opened the door.

Alex stood on the step, shoulders broad against the mist. His shirt was damp from sea spray; dark hair plastered across his brow. His hands, scarred from rope and salt, gripped a lantern that burned low, casting a dull amber glow around him. He had no business being there so late, and yet he looked like a man who had walked straight through the dark because nothing else could keep him away.

"You shouldn't be out," Olivia whispered, glancing toward the lane.

"If anyone sees,"

"Let them," he interrupted. His voice was low but firm. "You've been alone with this long enough."

Her throat tightened. She wanted to protest, but the words tangled.

He stepped inside before she could, shutting the door against the mist.

The sudden quiet between them felt heavier than the sea.

Alex's gaze flicked to her sleeve, to the faint glow bleeding through the linen. He didn't flinch. Instead, he set the lantern down and said softly, "Does it hurt?"

Olivia hesitated, then admitted, "Always."

Something passed over his face, anger, not at her but for her. "You take the sting, and the town takes the fear. That's not balance. That's sacrifice."

"It was my mistake," she said sharply, then softer, "my burden."

He shook his head. "You don't get to claim all the blame and all the pain, Olivia. Not when the sea wants more than one life."

Her breath caught. He had never spoken her name that way before, quiet, raw, like it mattered more than the whole village. She turned away quickly, afraid he might see how it shattered something inside her.

Behind her, her grandfather stirred by the fire. He watched them both through half-lidded eyes, unseen by Olivia, but Alex caught the old man's gaze and stiffened like a boy discovered sneaking sweets.

The old man's mouth tugged in the faintest wry smile. "If you're going to stand watch at her door, boy, at least bring her a fresh loaf from the market next time. A woman can't live on seaweed and guilt alone."

Olivia, still staring at her desk, didn't catch the meaning. She muttered nervously, "I have enough food," and reached for her journal.

Alex's ears reddened. He cleared his throat, muttering, "I'll… keep that in mind."

Her grandfather leaned back, satisfied, and let his eyes close again, the ghost of a chuckle escaping beneath his breath.

Olivia bent over the page once more. The line of her shoulders was taut, the glow beneath her sleeve throbbing faintly with every heartbeat. She dipped her pen, but her hand shook. Alex moved closer, steadying the inkpot before it tipped.

"Why are you really here?" she asked, barely above a whisper.

He met her gaze without wavering. "Because I don't want to watch you destroy yourself for this town. Because someone has to remind you that you're more than your mistakes."

The lamp sputtered, throwing their shadows long against the wall. For a heartbeat, the silence between them felt like a vow itself, unspoken, dangerous, inescapable.

Her grandfather's voice cut through, softer now, but steady as the tide. "You'll bear what comes next, child. But you won't bear it alone. That much, I'll see to."

Olivia closed her journal and met his gaze. Fear licked at her bones, but something harder settled inside her, the sharp line of resolve.

"If I cannot save Haven's Reach," she said, voice steady despite the weight in her chest, "then I will find a way to keep it from falling by myself. This isn't just our fight anymore. The sea carries further than these cliffs."

They stood in silence, listening to the distant pulse of light across the horizon. It spread wider than before, scattered like a chain of cold stars, a promise that the swarm was gathering, not retreating.

Olivia pressed her glowing arm to her chest and whispered, more to herself than to them both:

"This is only the beginning."

Alex reached for her hand then, not her wrist, not the bandaged place she hated, but her unmarked hand, warm and human and hers. He didn't speak, but his grip said what words would have broken: Whatever comes, I stand with you.

And though Olivia told herself she didn't notice, her grandfather, eyes still closed in the chair, smiled faintly and murmured, "About time." Alex had stayed the night.

Her grandfather had offered him the spare room as if it were nothing. "No sense sending a man out when the sea itself prowls at the windows," he'd said, setting an extra mug on the table like it was already decided. Alex had given only a curt nod, but his eyes had flicked briefly toward Olivia, steady, protective, and unreadable. Later, as he carried his boots upstairs, her grandfather muttered just loud enough for her to hear: "At least one lad in this town knows how to keep watch." Olivia, too tired to puzzle over his tone, missed the quiet smile tucked into his beard.

Before dawn, Olivia slipped out.

The sky was still black at its crown, but along the horizon a scarlet seam split the dark, spilling light across the restless sea. The cliffs were cold beneath her bare feet as she walked to the edge, the air raw with salt and brine.

She began to dance.

It was not a decision, not even a memory at first. It was instinct. Her body remembered what her mind had long buried, the dance her mother had shown her as a child, the dance meant to honor the ocean's breath and the fragile net of life within it. She lifted her arms in slow arcs, tracing the rise and fall of waves, turning in circles that marked the cycles of tide and moon. Her cloak whipped in the wind, her hair tangled around her face, but she kept moving, each step a prayer she hadn't known she still carried.

Her mind blurred, dazed with grief and longing. She saw her mother's smile, her father's callused hands steadying hers on the sand as she stumbled through the steps. For a heartbeat, she was no longer alone on the cliff.

Behind her, Alex had woken at the sound of the door. Some instinct pulled him after her, silent as a shadow, until he found her framed against the breaking horizon. He froze. She looked unearthly, not Olivia the weary scientist, but Olivia as the daughter of the sea, moving as though she belonged to its rhythm. He did not call her name. He could not. He simply watched.

Then the change came.

Her bandaged hand began to burn. The wrappings split with a faint hiss of light. From her palm, lines of luminescence unfurled, thin, trembling vines of brilliance that spilled into the air and cascaded downward, like a thousand fireflies loosed at once. They fell past the cliff's edge, raining into the sea below.

Where they touched, the black water softened. The scarlet reflection of dawn melted into true blue, rippling outward in rings. The tide pools along the rocks flickered, then steadied into the healthy glow of clear shallows. The color spread wider, as though the ocean itself breathed again.

Olivia's wrist burned fiercely. When she looked down, gasping, the old wound no longer pulsed with raw fire. Instead, a pattern had etched itself there, fine lines curling like tendrils, inked in living blue, a mark that looked less like a scar and more like a tattoo drawn by the sea itself.

Her breath shuddered. She hadn't noticed Alex step closer until his hand hovered near her arm, not touching, but steady. His eyes were wide, not with fear, but with awe.

"It's beautiful," he said softly, reverently.

Olivia blinked at him, at the way the dawn caught in his hair and the wind curled his shirt tight against his frame. She wanted to answer, but her throat locked. She turned back toward the ocean instead, her hand still trembling, the vines of light slowly fading into the waves.

Behind them, her grandfather lingered at the cliff path, cane braced against the stones. His weathered face was quiet with memory as he looked down at the girl-turned-woman who moved in the dawnlight.

"When she was small," he said at last, his voice carrying softly, "she'd come up here at sunrise, when the tide was low. Danced on this very cliff or down in the tide pools with her ankles bare in the water. Said the sea needed someone to remember its joy as much as its temper. Her mother taught her the steps, gentle turns, hands open to the horizon. I thought it was just a child's fancy."

He shook his head, a smile ghosting his lips, though his eyes glistened in the scarlet half-light. "Seems it wasn't fancy at all. The sea has been waiting for her to remember."

Alex stood a little apart, shoulders squared but eyes fastened wholly on Olivia. He said nothing, though the tightness in his jaw softened with each step of her dance, as if watching her unspooled something inside him too. The glow of her movements lit his features faintly, and for a heartbeat it seemed he belonged to this place as much as she did.

Olivia herself did not hear them. She was somewhere else, in a half-remembered moment of her mother's hand guiding hers, of laughter echoing across the waves, of her father's proud clapping from the cliff. Her body moved as if carried by that memory, each gesture linking past to present.

Then it happened.

The light in her wounded hand surged, no longer a wound but a wellspring. Vines of blue radiance spread from her fingers, spraying outward in arcs that struck the waves below. Where the light touched, the sea transformed, its sickly pallor vanishing, replaced with the clear, jeweled blue of unspoiled water. Foam glistened white, kelp shone green again, and the tide itself seemed to sigh with relief.

The glow faded, leaving the ocean rippling clean beneath the newborn sun. Olivia swayed once, catching herself on Alex's steadying arm, still lost in her daze.

Her grandfather's chuckle carried low, tinged with awe. "A child's dance," he murmured, half to himself. "And now the sea remembers her name."

Olivia lifted her hand at last, breath catching. The bandages had fallen away unnoticed in the wind. Where the burn had scarred her flesh, a pattern now remained, dark, intricate, and luminous all at once. It spiraled across her wrist like the tendrils of a jellyfish, etched as if the sea itself had claimed her.

She touched it with trembling fingers, whispering into the wind,

"This is only the beginning."